Editorial A

ELECTRICIAN'S GUIDE

TO THE

16th EDITION

OF THE

IEE WIRING REGULATIONS
BS 7671

THIRD EDITION
(revised January 1997 to include all
amendments and revisons to date)

JOHN WHITFIELD

Published by E·P·A Press
Wendens Ambo, Essex

Published by E·P·A Press, Wendens Ambo, UK

Whilst the author and the publishers have made every effort to ensure that the information and guidance given in this work is correct, all parties must rely upon their own skill and judgment when making use of it. Neither the author nor the publishers assume any liability to anyone for any loss or damage caused by any error or omission in the work, whether such error or omission is the result of negligence or any other cause. Any and all such liability is disclaimed.

The author and the publishers are grateful to the Institution of Electrical Engineers for permission to reproduce extracts from the 16th Edition of its Regulations for Electrical Installations. These Regulations are definitive and should always be consulted in their original form in case of doubt.

Additional copies of this book should be available through any good bookshop. In case of difficulty please contact the publishers directly at
E·P·A Press
Blythburgh House
Wendens Ambo
Saffron Walden, CB11 4JU

Tel 01799 541207, Fax 01799 541166

Pads of Completion, Inspection, Test and Measurement Certificates are available directly from the publishers at the above address.

British Library Cataloguing in Publication Data

Whitfield, J.F. (Frederic), *1930 -*

The electrician's guide to the 16th edition of the IEE wiring regulations
I. Title
621.31924
ISBN 0 9517362 6 4   3rd Edition
ISBN 0 9517362 4 8 (2nd Edition)
ISBN 0 9517362 1 3 (1st Edition)

Printed in the United Kingdom by St Edmundsbury Press

Contents

Contents

Preface to the Third Edition

The publication of the first amendments to BS 7671 (the 16th Edition of the IEE Wiring Regulations) in December, 1994, led to this third edition of the *'Electrician's Guide'*. Whilst the amendments are not far-reaching in themselves, the numerous changes in detail have been incorporated in the new edition. As well as the change in nominal supply voltage discussed below, changes have occurred to the Regulations for isolation and switching, PELV, swimming pools, agricultural and horticultural premises, and caravans and their sites.

The change in nominal supply voltages in Great Britain from 240/415 V to 230/400 V has not resulted in the complete reworking of exercises in the book which are based on the old voltage levels. This failure to change is a conscious act, based on the intention of the Electricity Supply Companies to take no immediate action. The actual voltages to be used in Great Britain will remain at 240/415 V for the foreseeable future, so the book continues to be based on these voltages. There may be problems in the future, and these are considered at the beginning of the book under the heading "Notes on supply voltage level".

The symbol V to denote voltage has been changed to U following the international adoption, but it should be noted that V remains the abbreviation for the volt. A list of all Regulations changed in the first amendment is appended in the hope that it will be helpful to electrical designers. The 15th Edition is now ancient history, so the chapter of the book outlining the changes to it appearing in the 16th Edition has been removed.

I should like to take this opportunity to thank the many readers who have contacted me to indicate their appreciation of the first two editions; I hope that they will find this edition equally rewarding.

John Whitfield
Norwich

By the end of December 1996 second editions of all six Guidance Notes had been published. They include important detail which is now incorporated in this revised edition. More than 100 pages have been changed and so the alterations are significant.

John Whitfield
Norwich
December 1996

The IEE Regulations, BS 7671
and this Guide

1.1   The need for this Electrician's Guide

The Institution of Electrical Engineers (IEE) has published an 'On-site guide'
with the 16th edition of its Regulations, which is intended to enable the
electrician to carry out 'certain specified installation work' without further
reference to the Regulations. Publicity for this guide before its issue stated
'the electrician is generally not required to perform any calculations'. When
printed, this was changed to *'to reduce the need for detailed calculations'*.

In the opinion of the Author, this attitude is *incorrect*. It assumes that
behind every electrician there is a designer who will provide him with precise details of exactly what he is to do. This is, of course, what happens in
some cases, but totally ignores all those electricians (probably a majority)
who are left to work entirely on their own, who have to do their own calculations and make their own decisions. These same electricians are subject to
the law, so that a failure to implement the IEE Regulations which leads to an
accident, may result (and has resulted) in a prison sentence.

It would be foolish to suppose that this *Electrician's Guide* could totally
replace the complete Regulations, which are made up of over two hundred
and fifty A4 pages, together with a total of eight associated 'guides'. Certainly, every electrician who does not have the advantage of expert design
advice should equip himself with the complete Regulations and with the
associated guides. However, it is the belief of the Author that this *Electrician's Guide* will help the average electrician to understand, and to implement, these very complicated Regulations in the safest and most cost-effec-
tive way possible.

Note on Supply Voltage Level

For many years the supply voltage for single-phase supplies in the UK
has been 240 V +/- 6%, giving a possible spread of voltage from 226 V
to 254 V.   For three-phase supplies the voltage was 415 V +/- 6%, the
spread being from 390 V to 440 V.   Most continental voltage levels
have been 220/380 V.

In 1988 an agreement was reached that voltage levels across Europe
should be unified at 230 V single phase and 400 V three-phase. In both cases
the tolerance levels have become -6% to +10%, giving a single-phase voltage spread of 216 V to 253 V, with three-phase values between 376 V and
440 V. In January 2003 the tolerance levels will be widened to +/- 10%.

Since the present supply voltages in the UK lie within the acceptable
spread of values, Supply Companies are not intending to reduce their voltages
in the near future. This is hardly surprising, because such action would im-
mediately reduce the energy used by consumers (and the income of the Companies) by more than 8%.

In view of the fact that there will be no change to the actual voltage
applied to installations, it has been decided not to make changes to the cal-

culations in this book. All are based on the 240/415 V supply voltages which have applied for many years and will continue so to do.

In due course, it is to be expected that manufacturers will supply appliances rated at 230 V for use in the UK. When they do so, there will be problems. A 230 V linear appliance used on a 240 V supply will take 4.3% more current and will consume almost 9% more energy. A 230 V rated 3 kW immersion heater, for example, will actually provide almost 3.27 kW when fed at 240 V. This means that the water will heat a little more quickly and that there is unlikely to be a serious problem other than that the life of the heater may be reduced, the level of reduction being difficult to quantify.

Life reduction is easier to specify in the case of filament lamps. A 230 V rated lamp used at 240 V will achieve only 55% of its rated life (it will fail after about 550 hours instead of the average of 1,000 hours) but will be brighter and will run much hotter, possibly leading to overheating problems in some luminaires. The starting current for large concentrations of discharge lamps will increase dramatically, especially when they are very cold. High pressure sodium and metal halide lamps will show a significant change in colour output when run at higher voltage than their rating, and rechargeable batteries in 230 V rated emergency lighting luminaires will overheat and suffer drastic life reductions when fed at 240 V.

There could be electrical installation problems here for the future!

1.2   The IEE Regulations

1.2.1   International basis

All electricians are aware of 'The Regs'. For well over one hundred years they have provided the rules which must be followed to make sure that electrical installations are safe.

A publication such as 'the Regs' must be regularly updated to take account of technical changes, and to allow for the 'internationalisation' of the Regulations. The ultimate aim is that all countries in the world will have the same wiring regulations. National differences make this still a dream, but we are moving slowly in that direction. The 15th Edition, when it was published in 1981, was the first edition of the IEE Regulations to follow IEC guidelines, and as such was novel in Great Britain. It was totally different from anything we had used before. The 16th Edition has not come as such a shock; certainly there are changes when we compare it with 'the 15th', but the two follow similar paths to reach the common destination of safety. Any electrician who has come to terms with the 15th should have little difficulty in working to the 16th.

A word is necessary about the identification of parts of this *Electrician's Guide* and of the IEE Regulations. In this *Electrician's Guide*, Regulation numbers are separated by a hyphen and indicated by placing them in square brackets. Thus, [514-06-03] is the third Regulation in the sixth group or subsection of Section 4 of Chapter 1 in Part 5. To avoid confusion, sections and sub-sections of this *Electrician's Guide* are divided by full points and enclosed in curly brackets. Hence, {5.4.6} is the sixth sub-section of section 4 of chapter 5 of this *Electrician's Guide*.

1.2.2   The Sixteenth Edition

This latest Edition of the IEE Wiring Regulations was published in May 1991. It was permissible to use the 15th Edition for installations put into service before January 1st, 1993. However, from now onwards the 16th Edition must be used. The current trend is to move towards a set of wiring regulations with worldwide application. IEC publication 364 *'Electrical Installations of Buildings'* has been available for some time, and the 16th Edition is based on many of its parts. The European Committee for Electrotechnical Standardisation (CENELEC) uses a similar pattern to IEC 364 and to the Wiring Regulations.

The introduction of the Free European Market in 1993 might well have caused serious problems for UK electrical contractors because whilst the IEE Wiring Regulations were held in high esteem, they had no legal status which would require Continentals who were carrying out installation work in the UK to abide by them. This difficulty was resolved in October 1992 when the IEE Wiring Regulations became a British Standard, BS 7671, giving them the required international standing.

It does not follow that an agreed part of IEC 364 will automatically become part of the IEE Wiring Regulations. For example, the IEE Wiring Regulations Committee is unable to agree with international rules which allow the installation of sockets in bathrooms, and so sockets are not allowed in these situations when we follow the 16th Edition.

BS 7671 recognises all harmonised standards (or Harmonised Documents, HDs) which have been agreed by all member states of the European Union. BS EN standards are harmonised standards based on harmonised documents and are published without addition to or deletion from the original HDs. When a BS EN is published the relevant BS is superseded and is withdrawn. A harmonised standard *eg* BS 7671, may have additions but not deletions, from the original standard. IEC and CENELEC publications follow the pattern which will be shown in {1.2.3}, and it is not always easy to find which Regulations apply to a given application. For example, if we need to find the requirements for bonding, there is no set of Regulations with that title to which we can turn. Instead, we need to consider five separate parts of the Regulations, which in this case are:

1   [Chapter 13]   Regulation [130-04-01],
2   [Section 413]   Regulations [413-02-15, 413-02-27 & 413-02-29],
3   [Section 514]   Regulation [514-13-01],
4   [Section 541]   complete, and
5   [Section 547]   complete.

The question arises 'how do we know where to look for all these different Regulations'? The answer is two-fold. First, the Regulations themselves have a good index. Second, this *Electrician's Guide* also has a useful index, from which the applicable sub-section can be found. At the top of each sub-section is a list of all applicable Regulations.

The detail applying to a particular set of circumstances is thus spread in a number of parts of the Regulations, and the overall picture can only be appreciated after considering all these separate pieces of information. This Guide is particularly useful in drawing all this information together.

1.2.3   Plan of the Sixteenth Edition
The regulations are in seven parts as shown in {Table 1.1}.

Table 1.1 Arrangement of the 16th Edition Parts	
Part 1	Scope, objects and fundamental requirements for safety
Part 2	Definitions
Part 3	Assessment of general characteristics
Part 4	Protection for safety
Part 5	Selection and erection of equipment
Part 6	Special installations or locations — particular requirements
Part 7	Inspection and testing

Also included in the Regulations are six Appendices, listed in {Table 1.2}. Unlike the 15th Edition, which was complete in itself, the 16th Edition has a number of publications called 'Guides' which include much material previously to be found in Appendices. These Guides must be considered to form part of the Regulations; their titles are shown in {Table 1.3}.

Table 1.2 Appendices to the 16th Edition	
Appendix 1	British Standards to which reference is made in The Regulations
Appendix 2	Statutory regulations and associated memoranda
Appendix 3	Time/current characteristics of overcurrent protective devices
Appendix 4	Current carrying capacities and voltage drop for cables and flexible cords
Appendix 5	Classification of external influences
Appendix 6	Forms of Completion and Inspection certificate

It is important to understand the relationship of the Appendices and of the Guidance Notes. Appendices provide information which the designer must have if his work is to comply with the Regulations. Other information, such as good practice, is contained in the Guidance Notes. Publication of the On-Site Guide, together with the first six of the Guidance Notes has resulted in minor changes and additions to *The Electricians' Guide*. This Edition includes these changes. Further changes will be made when the final Guidance Note is published by the IEE.

It is also important to have an understanding of the layout of the Regulations, so that work can be clearly identified. Each *Part* is divided into *Chapters*, which in turn are broken down into *Sections*, and then into *Groups* or *Sub-sections*, within which are to be found the actual regulations themselves. The particular regulation is identified by a number, such as 471-13-03. Note that this Regulation number is spoken as 'four seven one dash one three dash zero three' and NOT as 'four hundred and seventy one dash thirteen dash three'.

Table 1.3 Guides and Guidance Notes to the Regulations	
	On-site guide
1	Selection and erection
2	Isolation and switching
3	Inspection and testing
4	Protection against fire
5	Protection against electric shock
6	Protection against overcurrents
7	Special installations and locations

Note that whenever a group of digits is separated by a hyphen, the numbers represent a regulation; they are further identified in this Guide by placing them in square brackets, eg [471-13-03]. The apparent duplication of work within the Regulations may seem to be strange, but is necessary if the internationally agreed layout is to be followed.

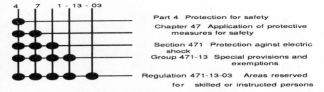

The international nature of the Regulations sometimes provides some strange reading. For example, regulations [413-02-21 to 413-02-26] deal with systems which are earthed using the IT method, described in {5.2.6} of this

Electrician's Guide. However, such a system is not accepted for public supplies in the UK and is seldom used except in conjunction with private generators.

1.3   The Electrician's Guide

1.3.1   The rationale for this Guide

In the first section of this chapter the reason for writing this Guide has been explained in general terms. Two statements in [120-01-02] point out a need for detailed guidance. They are:

1   *The Regulations are intended to be cited in their entirety if referred to in any contract.*
2   *They are not intended.......to instruct untrained persons.*

The first statement indicates the need for the Regulations to be written clearly and unambiguously, using language which will be understood by lawyers. If they were not, differences of opinion and interpretation would arise, which might not only lead to complicated law suits, but could possibly result in some unsafe installation work. This need for legal clarity does not always run parallel with technical clarity. The electrician, as well as his designer, if he has one, sometimes has difficulty in deciding exactly what a regulation requires, as well as the reasons for the regulation to be a requirement in the first place.

The second statement makes it clear that the legal wording of the Regulations is paramount, and that it will not be simplified to make its meaning more obvious to the electrician who is not expert in legal matters.

The object of this *Electrician's Guide* is to *clarify* the meaning of the Regulations and to *explain* the technical thinking behind them as far as is possible in a publication of this size. It is not possible to deal in detail with every Regulation. A choice has been made of those regulations which are considered to be most often met by the electrician in his everyday work. It must be appreciated that this *Electrician's Guide* cannot (and does not seek to) take the place of the Regulations, but merely to complement them. In some cases the 16th Edition provides detailed guidance, and where Tables contain such information its source is given. In other cases, where no guidance is otherwise provided, the Author makes suggestions in Tables which carry no details of the source.

1.3.2   Using this Electrician's Guide

A complete and systematic study of this *Electrician's Guide* should prove to be of great value to those who seek a fuller understanding of the Regulations. However, it is appreciated that perhaps a majority of readers will have neither the time nor the inclination to read in such wide detail, and will want to use the book for reference purposes. Having found the general area of work, using either the contents' list or the index of this *Electrician's Guide*, the text should help them to understand the subject. In the event of a regulation number being known, the cross reference index to this *Electrician's Guide* will show where an explanation can be found.

It has been previously stated that it is necessary for associated regulations to be spread out in various Parts and Chapters of the Wiring Regulations. In this *Electrician's Guide* they are collected together, so that, for example, all work on Earthing will be found in Chapter 5, on Testing and Inspection in Chapter 8, and so on. This *Electrician's Guide* is not a textbook of electrical principles. If difficulty is experienced in this respect, reference should be made to 'Electrical Craft Principles' by the same author, published by the IEE. This book has been written specifically for electricians.

Installation requirements and characteristics

2.1 Introduction

The first three parts of the Regulations lay the foundations for more detailed work later.

Part 1 **Scope, object and fundamental requirements for safety** applies to all installations and sets out their purposes as well as their status.

Part 2 **Definitions** details the precise meanings of terms used.

Part 3 **Assessment of general characteristics** is concerned with making sure that the installation will be fit for its intended purposes under all circumstances. The assessment must be made before the detailed installation design is started.

2.2 Safety requirements [Part 1]

2.2.1 Scope of the Regulations ~ [Chapter 11]

[Chapter 11] details exactly which installations are covered in the Regulations. The Regulations apply to all electrical installations except those relating to railway rolling stock, ships, offshore oil and gas rigs, aircraft, mines, quarries, radio interference equipment, lightning conductors other than bonding them to the electrical installation and those aspects of like installations covered by BS 5655. The 16th Edition includes, for the first time, installations associated with swimming pools, saunas, highway power supplies, street furniture and caravan parks and pitches.

Although the Regulations cover them, certain types of installation such as electric signs, emergency lighting, flameproof systems, fire detection and alarm systems, telecommunication systems and surface heating devices, are not fully covered. In such cases, the appropriate BS (British Standard) must be consulted as well as the Regulations. Equipment used as part of an installation but not constructed on site (a good example is switchgear) is not directly covered except that it is required to comply with the appropriate BS.

The Regulations apply to all installations fed at voltages included in the categories Extra-low-voltage and Low-voltage. The values of these classifications are shown in {Table 2.1}.

Table 2.1 Voltage classification

Extra-low-voltage —	not exceeding 50 V ac or 120 V dc
Low-voltage —	greater than extra-low voltage but not exceeding:-
	1000 V ac or 1500 V dc between conductors
	600 V ac or 900 V dc from any conductor to earth

2.2.2 Legal status of the Regulations [120-01 to 120-3, Appendix 2]

The Regulations are intended to provide safety to people or to livestock from fire, shock and burns in any installation which complies with their requirements. They are not intended to take the place of a detailed specification, but may form part of a contract for an electrical installation. The Regulations themselves contain the legal requirements for electrical installations, whilst the Guidance Notes indicate good practice.

In premises licensed for public performances, such as theatres, cinemas, discos and so on, the requirements of the licensing authority will apply in addition to the Regulations. In mine and quarry installations the requirements of the Health and Safety Commission must be followed and are mandatory.

In Scotland, the IEE Regulations are cited in the Building Regulations, so they must be followed. Whilst failure to comply with the Regulations has not generally been a criminal offence, those who complete such installations may be liable to prosecution in the event of an accident caused by the faulty wiring system.

The Electricity at Work Regulations 1989 became law on 1.1.92 in Northern Ireland and on 1.1.90 in the rest of the UK. Their original form made it clear that compliance with IEE Regulations was necessary, although it did not actually say so. The only buildings in which it may be argued that people are not at work are homes, so only domestic installations are *not* required to follow the requirements of the IEE Regulations, although even in these situations a prosecution may follow an accident. It may be helpful to mention '*The Guide to Electrical Safety at Work*' by John Whitfield, also published by EPA Press, which provides useful explanations of these Regulations.

The IEE Wiring Regulations became BS 7671 on 2nd October 1992 so that the legal enforcement of their requirements is easier, both in connection with the Electricity at Work Regulations and from an international point of view. The Construction (Design & Management) Regulations (CDM) were made under the Health & Safety at Work Act and implemented on 1.1.96. They require that installation owners and their designers consider health and safety requirements during the design and construction and throughout the life of an installation, including maintenance, repair and demolition.

2.2.3 *New inventions or methods* ~ [120-4 and 120-05]

The Regulations indicate clearly that it is intended that only systems and equipments covered by the Regulations should be used. However, in the event of new inventions or methods being applied, these should give at least the same degree of safety as older methods which comply directly.

Where such new equipments or methods are used, they should be noted on the Completion Certificate (*see* {8.8.2}) as not complying directly with the requirements of the Regulations.

2.2.4 *Safety requirements* ~ [Chapter 13]

Safety is the basic reason for the existence of the Wiring Regulations. [Chapter 13] has the title 'Fundamental Requirements for Safety' and really contains, in shortened form, the full safety requirements for electrical installations. It has been said that the twenty short regulations in Chapter 13 *are* the Regulations and that the rest of the publication simply serves to spell out their requirements in greater detail. For example, [512-04-01], part of the Common Rules for the Selection and Erection of Equipment, has precisely the same meaning as [130-02-02] in requiring that the installation should be capable of carrying the maximum power required by the system when it is functioning in the way intended. It is important to appreciate that the Electricity at Work Regulations apply to all electrical installations, covering designers, installers, inspectors, testers and users. The regular inspection and testing of all electrical installations is a requirement of the Electricity at Work Regulations.

Perhaps the most basic rule of all, [130-2-02], is so important that it should be quoted in full. It states:-

> Good workmanship and proper materials shall be used

The details of [Chapter 13] are covered more fully later in this Guide.

The Smoke Detector Act 1991 requires that all new dwellings must be fitted with mains-fed smoke detectors to BS 5446 Part 1 but as of January 1997 no date of implementation had been announced.

2.2.5   Supplies for safety services   ~   [56]

Safety services are special installations which come into use in an emergency, to protect from, or to warn of, danger and to allow people to escape. Thus, such installations would include fire alarms and emergency lighting, supplies for sprinkler system pumps, as well as specially protected circuits to allow lifts to function in the event of fire.

The special needs of safety circuits will often be required by authorities other than the IEE Regulations, especially where people gather in large numbers. For example, safety circuits in cinemas are covered by the Cinematograph Regulations 1955, administered by the Home Office in England, Wales and Northern Ireland and in Scotland by the Secretary of State.

Safety circuits cannot be supplied by the normal installation, because it may fail in the dangerous circumstances the systems are there to guard against. The permitted sources of supply include cells and batteries, standby generators and separate feeders from the mains supply. The latter must only be used if it is certain that they will not fail at the same time as the main supply source.

[Chapter 56] contains six sections and a total of twenty-one Regulations, detailing the requirements for safety services. In effect, the circuits concerned must comply with all of the rest of the Regulations, and with some additional needs.

The safety source must have adequate duration. This means, for example, that battery operated emergency lighting must stay on for the time specified in the applicable British Standard (BS 5266). Since such installations may be called on to operate during a fire, they must not be installed so that they pass through fire risk areas and must have fire protection of adequate duration.

Safety circuits must be installed so that they are not affected by faults in normal systems and overload protection can be omitted to make the circuits less liable to failure. Safety sources must be in positions which are only open to skilled or instructed persons, and switchgear, control gear and alarms must be suitably labelled to make them clearly identifiable.

2.3   Definitions [Part 2]

Any technical publication must make sure that its readers are in no doubt about exactly what it says. Thus, the meanings of the terms used must be defined to make them absolutely clear. For this purpose, [Part 2] includes about one hundred and forty-five definitions of words used. For example, the term 'low voltage' is often assumed to mean a safe level of potential difference. As far as the Regulations are concerned, a low voltage could be up to 1000 V ac or 1500 V dc!

Important definitions are those for skilled and instructed persons. A *skilled person* is one with technical knowledge or experience to enable the avoidance of dangers that may be associated with electrical energy dissipation. An *instructed person* is one who is adequately advised or supervised by skilled persons to enable dangers to be avoided.

The reader of the Regulations should always consult [Part 2] if in doubt, or even if he suspects that there could possibly be a different meaning from the one assumed. If the word or phrase concerned is not included in [Part 2], BS 4727, 'Glossary of electrotechnical, power, telecommunication, electronics, lighting and colour terms' should be consulted.

2.4   Assessment of general characteristics [Part 3]

2.4.1   Introduction   ~   [300-01]

Before planning and carrying out an electrical installation, thought must be given to factors affecting its effective operation, but some basic information will be needed concerning the supply system. The Electricity Supply Company will provide information concerning the number of phases, phase sequence and voltage of the supply. They will also give information as to the prospective short circuit current {3.7.2}. Details of the supply protection and the external earth fault loop impedance {5.3} for the site concerned will be given.

The installation designer will also need to consider other matters, such as the purpose of the installation, the external influences which may affect it {2.4.3}, the compatibility of its equipment {2.4.4} and its maintainability {2.4.5}.

2.4.2   *Purposes, supplies and structure*   ~ [31, 512-01 to 512-04]
[Chapter 31] includes regulations which effectively provide a check list for the designer to ensure that all the relevant factors have been taken into account.

An abbreviated check list would include the items shown in {Table 2.2}. Where equipments or systems have different types of supply, such as voltage or frequency, they must be separated so that one does not influence the other.

Table 2.2   Initial check list for the designer

a)   the maximum demand
b)   the number and type of live conductors
c)   the type of earthing arrangement
d)   the nature of supply, including:
   i)   nominal voltage(s)
   ii)   nature of current (a.c. or d.c.) and frequency
   iii)   prospective short circuit current at the supply intake
   iv)   earth fault loop impedance of the supply to the intake position
   v)   suitability of the installation for its required purposes, including maximum demand
   vi)   type and rating of the main protective device (supply main fuse)
e)   the supplies for safety services and standby purposes
f)   the arrangement of installation circuits.

2.4.3   *External influences* ~ [32, 512-06, 515-01, Appendix 5]
This subject is a very clear example of the way that the IEE Wiring Regulations follow an international pattern. As no final agreement on the subject of external influences had been reached when the Sixteenth Edition was published, [Chapter 32] was effectively missing from the edition. It will be provided by an amendment when an agreement is reached internationally which also meets the requirements of the UK. A great deal of extra data on external influences has now (January 1997) been published in Appendix C of the second edition of Guidance Note 1.

[Appendix 5] gives a list of how external influences are to be indicated in three categories, A, B and C. The second letter refers to the nature of the external influence. These letters are different for each of the categories, and are shown in {Table 2.3}. Next follows a number, rising in some cases to 8, which indicates the class within each external influence. For example
BA5 indicates: B = utilisation
                             A = capability
                             5 = skilled persons
The ability of an enclosure to withstand the ingress of solid objects and of water is indicated by the index of protection (IP) system of classification. The system is detailed in BS EN 60529, and consists of the letters IP followed by two numbers. The first number indicates the degree of protection against solid objects, and the second against water. If, as is sometimes the case, either form of protection is not classified, the number is replaced with X. Thus, IPX5 indicates an enclosure whose protection against solid objects is not classified, but which will protect against water jets.

Table 2.3   External influences

Environment (A)

	AA	ambient temperature
	AB	humidity
	AC	altitude
	AD	water
	AE	foreign bodies
	AF	corrosion
	AG	impact
	AH	vibration
	AJ	other mechanical stresses
	AK	flora (plants)
	AL	fauna (animals)
	AM	radiation
	AN	solar (sunlight)
	AP	seismic (earthquakes)
	AQ	lightning
	AR	wind

Utilisation (B)

	BA	capability (such as physical handicap)
	BB	resistance
	BC	contact with earth
	BD	evacuation (such as difficult exit conditions)
	BE	materials (fire risk)

Building (C)

	CA	materials (combustible or nonflammable)
	CB	structure (spread of fire *etc*.)

Other letters are also used as follows

W	placed after IP indicates a specified degree of weather protection
S	after the numbers indicates that the enclosure has been tested against water penetration when not in use
M	after the numbers indicates that the enclosure has been tested against water penetration when in use.

A more complicated system involving the use of additional and of supplementary letters has been adopted internationally, as has an impact protection code (the IK Code). These details are beyond the scope of this Guide but full details are found in Appendix B of the 2nd Ed. of Guidance Note 1 An abbreviated form of the information concerning the meanings of the two numbers of the IP system is shown in {Table 2.4}.

Table 2.4   Numbers in the I P system

First number	Mechanical protection against	Second number	Water protection against
0	Not protected	0	Not protected
1	Solid objects exceeding 50 mm	1	Dripping water
2	Solid objects exceeding 12 mm	2	Dripping water when tilted up to 15°
3	Solid objects exceeding 2.5 mm	3	Spraying water
4	Solid objects exceeding 1.0 mm	4	Splashing water
5	Dust protected	5	Water jets
6	Dust tight	6	Heavy seas
		7	Effects of immersion
		8	Submersion

The EMC (Electromagnetic Compatibility) Regulations are now in force and require that electrical installations are designed and constructed so that they do not cause electromagnetic interference with other equipments or systems and are themselves immune to electromagnetic interference from other systems. The full implications of these Regulations for electrical installations are not yet fully understood.

2.4.4   *Compatibility* ~ [331-01, 512-05]

One part of an electrical installation must not produce effects which are harmful to another part. For example, the heavy transient starting current of electric motors {7.15.1} may result in large voltage reductions which can affect the operation of filament and discharge lamps. Again, the use of some types of controlled rectifier will introduce harmonics which may spread through the installation and upset the operation of devices such as electronic timers. Computers are likely to be affected by the line disturbances produced by welding equipment fed from the same system. 'Noisy' supplies, which contain irregular voltage patterns, can be produced by a number of equipments such as machines and thermostats. Such effects can result in the loss of data from computers, point-of-sale terminals, electronic office equipment, data transmission systems, and so on. Separate circuits may be necessary to prevent these problems from arising, together with the provision of 'clean earth' systems.

Very strict European laws limiting the amount of electromagnetic radiation permitted from electrical installations and appliances apply from January 1st, 1996 (the Electromagnetic Compatibility, or EMC, Directive). Equipment and installations must:
1. not generate excessive electromagnetic disturbances that could interfere with other equipments (such as radio sets), and
2. have adequate immunity to electromagnetic disturbances to allow proper operation in its normal environment.
(For further details see *The Guide to the EMC Directive 89/336/EEC, Second Edition,* by Chris Marshman also published by E·P·A Press.)

2.4.5   *Maintainability* ~ [130-02-01, 341-01, 513-01, 529]

When designing and installing an electrical system it is important to assess how often maintenance will be carried out and how effective it will be. For example, a factory with a staff of fully trained and expertly managed electricians who carry out a system of planned maintenance may well be responsible for a different type of installation from that in a small motor-car repair works which does not employ a specialist electrician and would not consider calling one except in an emergency.

In the former case there may be no need to store spare cartridge fuses at each distribution board. The absence of such spares in the latter case may well lead to the dangerous misuse of the protective system as untrained personnel try to keep their electrical system working.

The electrical installation must be installed so that it can be easily and safely maintained, is always accessible for additional installation work, maintenance and operation and so that the built-in protective devices will always provide the expected degree of safety. Access to lighting systems is a common problem, and unless the luminaires can be reached from a stable and level surface, or from steps of a reasonable height, consideration should be given to the provision of hoisting equipment to enable the lighting system to be lowered, or the electrician raised safely to the luminaires.

2.5   Standards ~ [510-01, 511-01, Appendix 1]

Every item of equipment which forms part of an electrical installation must be designed and manufactured so that it will be safe to use. To this end all equipment should, wherever possible, comply with the relevant British Standard; a list of the standards applying to electrical installations is included in the Regulations as [Appendix 1]. Increasingly we are meeting equipment which has been produced to the standards of another country. It is the responsibility of the designer to check that such a standard does not differ from the British Standard to such an extent that safety would be reduced. If the electrician is also the designer, and this may well happen in some cases, the electrician carries this responsibility.

Foreign standards may well be acceptable, but it is the responsibility of the designer to ensure that this is so. In the event of equipment not complying with the BS concerned, it is the responsibility of the installer to ensure that it provides the same degree of safety as that established by compliance with the Regulations.

A harmonised documant (HD) is one agreed by all member nations of CENELEC. These documents may be published in individual countries with additions (BS 7671 is one such) but not with deletions. BS EN documents are Standards agreed by all CENELEC members and published without additions or deletions. Such EN standards supersede the original British Standard.

An increasing number of electrical contractors are registered as having satisfied the British Standards Institution that they comply fully with BS 5750 — Quality Assurance. This standard is not concerned directly with the Wiring Regulations, but since compliance with them is an important part of the organisation's assurance that its work is of the highest quality, registration under BS 5750 implies that the Wiring Regulations are followed totally.

2.6   Undervoltage ~ [451-01, 535-01]

This chapter of the Regulations deals with the prevention of dangers that could occur if voltage falls to a level too low for safe operation of plant and protective devices. Another problem covered is the danger that may arise when voltage is suddenly restored to a system which has previously been on a lower voltage or without voltage at all. For example, a machine which has stopped due to voltage falling to a low level may be dangerous if it restarts suddenly and unexpectedly when full voltage returns. A motor starter with built-in undervoltage protection will be explained in {7.15.1}.

The attention of the installer and the designer is drawn to the possibility that low voltage may cause equipment damage. Should such damage occur, it must not cause danger. Where equipment is capable of operating safely at low voltage for a short time, a time delay may be used to prevent switching off at once when undervoltage occurs. This system may prevent plant stoppages due to very short time voltage failures. However, such a delay must not prevent the immediate operation of protective systems.

Installation control and protection

3.1   Introduction

Electrical installations must be protected from the effects of short circuit and overload. In addition, the people using the installations, as well as the buildings containing them, must be protected from the effects of fire and of other hazards arising from faults or from misuse.

Not only must automatic fault protection of this kind be provided, but an installation must also have switching and isolation which can be used to control it in normal operation, in the event of emergency, and when maintenance is necesssary.

This Chapter will consider those regulations which deal with the disconnection of circuits, by both manual and automatic means, the latter in the event of shock, short circuit or overload. It does not include the Regulations which concern automatic disconnection in the event of an earth fault: these are considered in {Chapter 5}.

In order that anyone operating or testing the installation has full information concerning it, a diagram or chart must be provided at the mains position showing the number of points and the size and type of cables for each circuit, the method of providing protection from direct contact (*see* {3.4.5}) and details of any circuit in which there is equipment, such as passive infra-red detectors or electronic fluorescent starters, vulnerable to the high voltage used for insulation testing.

3.2   Switching

3.2.1   *Switch positions* ~ [130-05-02, 460-01, 461-01 to 461-03, 476-01, 476-02-01, 530-01, 537-01 and 551-01-03]

A switch is defined as a device which is capable of making or breaking a circuit under normal and under overload conditions. It can make, but will not necessarily break, a short circuit, which should be broken by the overload protecting fuse or circuit breaker. A switching device may be marked with ON and OFF positions, or increasingly, the numbers 1 for ON and 0 for OFF are being used.

A semiconductor device is often used for switching some lighting and heating circuits, but will not be suitable for disconnecting overloads; thus, it must be backed up by a mechanical switch. The semiconductor is a functional switch but must NOT be used as an isolator.

{Figure 3.1} shows which poles of the supply need to be broken by the controlling switches. For the TN-S system (earth terminal provided by the Electricity Company), the TN-C-S system (protective multiple earthing) and the TT system (no earth provided at the supply), all phase conductors MUST be switched, but NOT the protective (earth) conductor.

The neutral conductor need not be broken except for:

1   the main switch in a single-phase installation, or
2   heating appliances where the element can be touched, or
3   autotransformers (not exceeding 1.5 kV) feeding discharge lamps

The neutral will need to be disconnected for periodic testing, and provision must be made for this; it is important that the means of disconnection is accessible and can only be completed with the use of a tool.

The protective conductor should never be switched, except when the supply can be taken from either of two sources with earth systems which must not be connected together. In this case the switches needed in the protective conductors must be linked to the phase switches so that it is impossible for the supply to be provided unless the earthing connection is present.

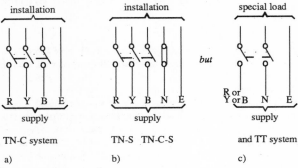

combined earth
and neutral (PEN)

installation   installation   special load

but

R   Y   B   E   R   Y   B   N   E   R or Y or B   N   E

supply   supply   supply

TN-C system   TN-S   TN-C-S   and TT system

a)   b)   c)

Fig 3.1 Supply system broken by switches
a) TN-C systems   b) TN-S, TN-C-S   c) TT systems

Every circuit must be provided with a switching system so that it can be interrupted on load. In practice, this does not mean a switch controlling each separate circuit; provided that loads are controlled by switches, a number of circuits may be under the overall control of one main switch. An example is the consumer unit used in the typical house, where there is usually only one main switch to control all the circuits, which are provided with individual switches to operate separate lights, heaters, and so on. If an installation is supplied from more than one source there must be a separate main switch for each source, and each must be clearly marked to warn the person switching off the supplies that more than one switch needs to be operated.

It should be noted that a residual current device (RCD) may be used as a switch provided that its rated breaking capacity is high enough.

3.2.2 Emergency switching ~ [130-06, 463-01, 476-03, 514-01 & 537-04]

Emergency switching is defined as rapidly cutting the supply to remove hazards. For example, if someone is in the process of receiving an electric shock, the first action of a rescuer should be to remove the supply by operating the emergency switch, which may well be the main switch. Such switching must be available for all installations. Note that if there is more than one source of supply a number of main switches may need to be opened (*see* 3.2.1). The designer must identify all possible dangers, including electric shock, mechanical movement, excessive heat or cold and radiation dangers, such as those from lasers or X-rays.

In the special case of electric motors, the emergency switching must be adjacent to the motor. In practice, such switching may take the form of a starter fitted close to the motor, or an adjacent stop button (within 2 m) where the starter is remote. Where a starter or contactor is used as an emergency switch, a positive means must be employed to make sure that the installation is safe. For example, operation should be when the operating coil is de-energised, so that an open circuit in the coil or in its operating circuit will cause the system to be switched off {Fig 3.2}. This is often called the 'fail-safe' system.

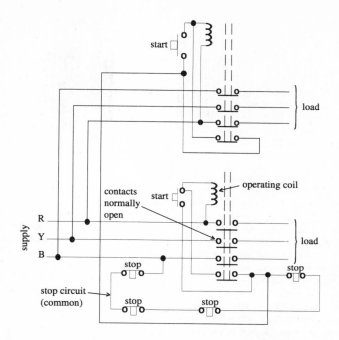

Fig 3.2   Two circuit breakers linked to a common stop circuit. The system is 'fail-safe'

To prevent unexpected restarting of rotating machines, the 'latching off ' stop button shown in {Fig 3.3} is sometimes used. On operation, the button locks (latches) in the off position until a positive action is taken to release it.

In single-phase systems, it must be remembered that the neutral is earthed. This means that if the stop buttons are connected directly to the neutral, a single earth fault on the stop button circuit would leave the operating coil permanently fed and prevent the safety system from being effective. It is thus essential for the operating coil to be directly connected to the neutral, and the stop buttons to the phase. Such an earth fault would then operate the protective device and make the system safe.

The means of emergency switching must be such that a single direct action is required to operate it. The switch must be readily accessible and clearly marked in a way that will be durable. Consideration must be given to the intended use of the premises in which the switch is installed to make sure as far as possible that the switching system is always easy to reach and to use. For example, the switch should not be situated at the back of a cupboard which, in use, is likely to be filled with materials making it impossible to reach the switch.

In cases where operation could cause danger to other people (an ex-ample is where lighting is switched off by operating the emergency switch), the switch must be available only for operation by instructed persons. Every fixed or stationary appliance must be provided with a means of switching which can be used in an emergency. If the device is supplied by an unswitched plug and socket, withdrawal of the plug is NOT acceptable to comply with this requirement; such action is acceptable for functional switching {3.2.4}.

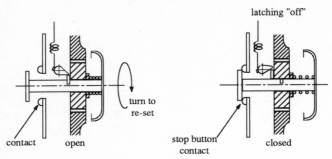

Fig 3.3 *'Latching-off' stop button*

Where any circuit operates at a p.d. (potential difference) exceeding low voltage a fireman's emergency switch must be provided. Such installations usually take the form of discharge lighting (neon signs), and this requirement applies for all external systems as well as internal signs which operate unattended. The purpose is to ensure the safety of fire fighters who may, if a higher voltage system is still energised, receive dangerous shocks when they play a water jet onto it. The fireman's switch is not required for portable signs consuming 100 W or less which are supplied *via* an easily accessible plug and socket.

The fireman's switch must meet the following requirements

1.   The switch must be mounted in a conspicuous position not more than 2.75 m from the ground.

2.   It must be coloured red and have a label in lettering at least 13 mm high 'FIREMAN'S SWITCH'. On and off positions should be clearly marked, and the OFF position should be at the top. A lock or catch should be provided to prevent accidental reclosure.

3.   For exterior installations the switch should be close to the load, or to a notice in such a position to indicate clearly the position of the well-identified switch.

4.   For interior installations, the switch should be at the main entrance to the building.

5.   Ideally, no more than one internal and one external switch must be provided. Where more become necessary, each switch must be clearly marked to indicate exactly which parts of the installation it controls.

6.   Where the local fire authority has additional requirements, these must be followed.

7   The switch should be arranged on the supply side of the step-up sign transformer.

3.2.3   *Switching for mechanical maintenance* [462-01 & 537-03]

Mechanical maintenance is taken as meaning the replacement and repair of non-electrical parts of equipment, as well as the cleaning and replacement of lamps. Thus, we are considering the means of making safe electrical equipment which is to be worked on by non-electrical people.

Such switches must be:

1.   easily and clearly identified by those who need to use them

2.   arranged so that there can be no doubt when they are on or off

3.   close to the circuits or equipments they switch

4.   able to switch off full load current for the circuit concerned

5.   arranged so that it is impossible for them to be reclosed unintentionally.

Where mechanical maintenance requires access to the interior of equipment where live parts could be exposed, special means of isolation are essential.

3.2.4 *Functional switching* ~ [464, 537-05]

Functional switching is used in the normal operation of circuits to switch them on and off. A switch must be provided for each part of a circuit which may need to be controlled separately from other parts of the installation. Such switches are needed for equipment operation, and a group of circuits can sometimes be under the control of a single switch unless separate switches are required for reasons of safety. Semiconductor switches may control the current without opening the poles of the supply. Fuses and links must not be used as a method of switching.

A plug and socket may be used as a functional switching device provided its rating does not exceed 16 A. It will be appreciated that such a device must not be used for emergency switching (*see* {3.2.2}).

3.3 Isolation

3.3.1 *Isolator definition* ~ [476-01, 537-02-01]

An isolator is not the same as a switch. It should only be opened when *not* carrying current, and has the purpose of ensuring that a circuit cannot become live whilst it is out of service for maintenance or cleaning. The isolator must break all live supply conductors; thus both phase and neutral conductors must be isolated. It must, however, be remembered that switching off for mechanical maintenance (*see* {3.2.3}) is likely to be carried out by non-electrically skilled persons and that they may therefore unwisely use isolators as on-load switches. To prevent an isolator, which is part of a circuit where a circuit breaker is used for switching, from being used to break load current, it must be interlocked to ensure operation only after the circuit breaker is already open. In many cases an isolator can be used to make safe a particular piece of apparatus whilst those around it are still operating normally.

3.3.2 *Isolator situation* ~ [461-01-01, 476-02-02 and 476-02- 03, 537-02-02 and 537-02-04 to 537-02-08]

Isolators are required for all circuits and for all equipments, and must be adjacent to the system protected.

As for a switch, the isolator must be arranged so that it will not close unintentionally due to vibration or other mechanical effect. It must also, where remote from the circuit protected, be provided with a means of locking in the OFF position to ensure that it is not reclosed whilst the circuit is being worked on. The isolator should also be provided, where necessary, with means to prevent reclosure in these circumstances. This may be a padlock system to enable the device to be locked in the OFF position, a handle which is removable in the off position only, *etc.*

It is particularly important for motors and their starters to be provided with adjacent isolation to enable safe maintenance.

The special requirements for isolating discharge lamp circuits operating at voltages higher than low voltage will be considered in {7.12.2}.

3.3.3 *Isolator positions* ~ [461-01-01, 461-01-04 and 530-02-04]

Every circuit must have its means of isolation, which must be lockable in the OFF position where remote from the equipment protected. The OFF position must be clearly marked in all cases so that there is no doubt in the mind of the operator as to whether his circuit is isolated and thus safe to work on. For single phase systems, both live conductors (phase and neutral) must be broken by the isolator (TT system). On three phase supplies, only the three phases (R, Y, and B or L1, L2 and L3) need to be broken, the neutral being left solidly connected (TN-S and TN-C-S systems).

This neutral connection is usually through a link which can be removed for testing. Clearly, it is of great importance that the link is not removed during normal operation, so it must comply with one or both of the following requirements:
1. it can only be removed by using tools, and/or
2. it is accessible only to skilled persons.

In many circuits capacitors are connected across the load. There are many reasons for this, the most usual in industrial circuits being power factor correction. When the supply to such a circuit is switched off, the capacitor will often remain charged for a significant period, so that the isolated circuit may be able to deliver a severe shock to anyone touching it.

The Regulations require that a means of discharging such capacitors should be provided. This usually takes the form of discharge resistors {Fig 3.4}, which provide a path for discharge current. These resistors are connected directly across the supply, so give rise to a leakage current between live conductors. This current is reduced by using larger resistance values, but this increases the time taken for the capacitors to discharge to a safe potential difference. In practice, a happy medium is struck between these conflicting requirements, resistors with values of about 100 kΩ being common.

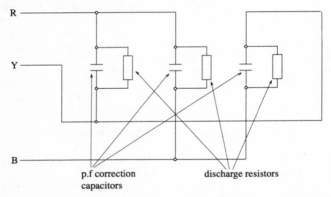

Fig 3.4   Discharge resistors connected to a three-phase capacitor bank

3.3.4   *Semiconductor isolators* ~ [537-02-03 and 537-05-02]
Semiconductors are very widely used in electrical installations, from simple domestic light dimmers (usually using triacs) to complex speed controllers for three phase motors (using thyristors). Whilst the semiconductors themselves are functional switches, operating very rapidly to control the circuit voltage, they must NOT be used as isolators.

This is because when not conducting (in the OFF position) they still allow a very small leakage current to flow, and have not totally isolated the circuit they control. {Figure 3.5} shows semiconductors in a typical speed control circuit.

3.3.5   *Isolator identification* ~ [461-01-02, 461-01-03, 461-01-05, 537-02-04 and 537-02-09]
The OFF position on all isolators must be clearly marked and should not be indicated until the contacts have opened to their full extent to give reliable

isolation. Every isolator must be clearly and durably marked to indicate the circuit or equipment it protects. If a single isolator will not cut off the supply from internal parts of an enclosure, it must be labelled to draw attention to the possible danger. Where the unit concerned is suitable only for off-load isolation, this should be clearly indicated by marking the isolator "Do NOT open under load".

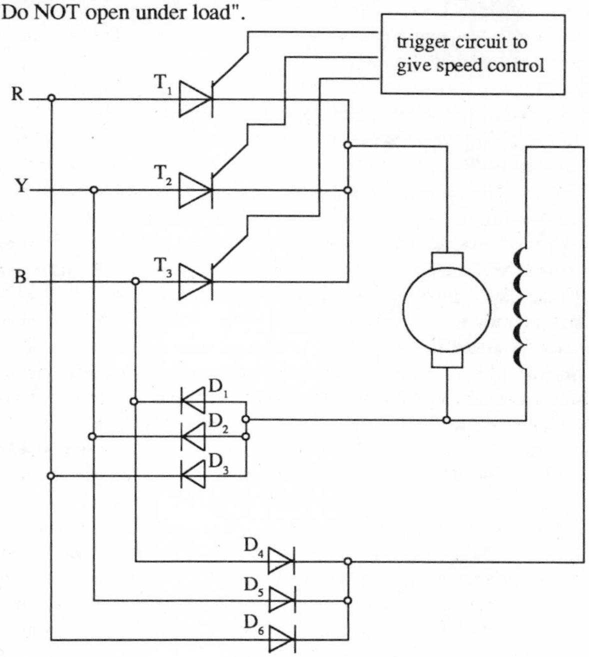

Fig 3.5   Speed control for a dc motor fed from a three-phase supply

3.4   Electric shock protection

3.4.1   *The nature of electric shock*

The nervous system of the human body controls all its movements, both conscious and unconscious. The system carries electrical signals between the brain and the muscles, which are thus stimulated into action.

The signals are electro-chemical in nature, with levels of a few millivolts, so when the human body becomes part of a much more powerful external circuit, its normal operations are swamped by the outside signals. The current forced through the nervous system of the body by external voltage *is* electric shock.

All the muscles affected receive much stronger signals than those they normally get and operate very much more violently as a result. This causes uncontrolled movements and pain. Even a patient who is still conscious is usually quite unable to counter the effects of the shock, because the signals from his brain, which try to offset the effects of the shock currents, are lost in the strength of the imposed signals.

A good example is the 'no-let-go' effect. Here, a person touches a conductor which sends shock currents through his hand. The muscles respond by closing the fingers on the conductor, so it is tightly grasped. The person wants to release the conductor, which is causing pain, but the electrical signals from his brain are swamped by the shock current, and he is unable to let go of the offending conductor.

The effects of an electric shock vary considerably depending on the current imposed on the nervous system, and the path taken through the body. The subject is very complex but it has become clear that the damage done to the human body depends on two factors:

1. the value of shock current flowing, and
2. the time for which it flows.

These two factors have governed the international movement towards making electrical installations safer.

3.4.2   *Resistance of the shock path*   ~   [400 and 471-01]

In simple terms the human body can be considered as a circuit through which an applied potential difference will drive a current. As we know from Ohm's Law, the current flowing will depend on the voltage applied and the resistance of the current path. Of course, we should try to prevent or to limit shock by aiming to stop a dangerous potential difference from being applied across the body. However, we have to accept that there are times when this is impossible, so the important factor becomes the resistance of the current path.

The human body is composed largely of water, and has very low resistance. The skin, however, has very high resistance, the value depending on its nature, on the possible presence of water, and on whether it has become burned. Thus, most of the resistance to the passage of current through the human body is at the points of entry and exit through the skin. A person with naturally hard and dry skin will offer much higher resistance to shock current than one with soft and moist skin; the skin resistance becomes very low if it has been burned, because of the presence of conducting particles of carbon.

In fact, the current is limited by the impedance of the human body, which includes self capacitance as well as resistance. The impedance values are very difficult to predict, since they depend on a variety of factors including applied voltage, current level and duration, the area of contact with the live system, the pressure of the contact, the condition of the skin, the ambient and the body temperatures, and so on.

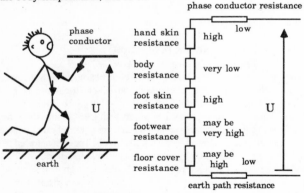

Fig 3.6   Path of electric shock current

{Figure 3.6} is a simplified representation of the shock path through the body, with an equivalent circuit which indicates the components of the resistance concerned. It must be appreciated that the diagram is very approximate; the flow of current through the body will, for example, cause the victim to sweat, reducing the resistance of the skin very quickly after the shock commences. Fortunately, people using electrical installations rarely have bare feet, and so the resistance of the footwear, as well as of the floor coverings, will often increase overall shock path resistance and reduce shock current to a safer level.

There are few reliable figures for shock current effects, because they differ from person to person, and for a particular person, with time. However, we know that something over one milliampere of current in the body produces the sensation of shock, and that one hundred milliamperes is likely quickly to prove fatal, particularly if it passes through the heart.

If a shock persists, its effects are likely to prove to be more dangerous. For example, a shock current of 500 mA may have no lasting ill effects if its duration is less than 20 ms, but 50 mA for 10 s could well prove to be fatal. The effects of the shock will vary, but the most dangerous results are ventricular fibrillation (where the heart beat sequence is disrupted) and compression of the chest, resulting in a failure to breathe.

The resistance of the shock path is of crucial importance. The Regulations insist on special measures where shock hazard is increased by a reduction in body resistance and good contact of the body with earth potential. Such situations include locations containing bath tubs or showers, swimming pools, saunas and so on. The Regulations applying to these special installations are considered in {Chapter 7}.

Another important factor to limit the severity of electric shock is the limitation of earth fault loop impedance. Whilst this impedance adds to that of the body to reduce shock current, the real purpose of the requirement is to allow enough current to flow to operate the protective device and thus to cut off the shock current altogether quickly enough to prevent death from shock.

How quickly this must take place depends on the level of body resistance expected. Where sockets are concerned, the portable appliances fed by them are likely to be grasped firmly by the user so that the contact resistance is lower. Thus, disconnection within 0.4 s is required. In the case of circuits feeding fixed equipment, where contact resistance is likely to be higher, the supply must be removed within 5 s. For situations where earth contact is likely to be good, such as farms and construction sites, disconnection is required within 0.2 s. Earth fault loop impedance is considered more fully in {Chapter 5}.

3.4.3 *Contact with live conductors* [413, 414, 470 and 471-01]

In order for someone to get an electric shock he or she must come into contact with a live conductor. Two types of contact are classified.

1 Direct contact

An electric shock results from contact with a conductor which forms part of a circuit and would be expected to be live. A typical example would be if someone removed the plate from a switch and touched the live conductors inside (*see* {Fig 3.7}). Overcurrent protective systems will offer no protection in this case, but it is possible that an RCD with an operating current of 30 mA or less may do so.

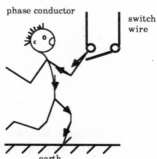

phase conductor

switch wire

Fig 3.7   *Direct contact*

earth

2   *Indirect contact.*

An electric shock is received from contact with something connected with the electrical installation which would not normally be expected to be live, but has become so as the result of a fault. This would be termed an exposed conductive part. Alternatively, a shock may be received from a conducting part which is totally unconnected with the electrical installation, but which has become live as the result of a fault. Such a part would be called an extraneous conductive part.

An example illustrating both types of indirect contact is shown in {Fig 3.8}. Danger in this situation results from the presence of a phase to earth fault on the kettle. This makes the kettle case live, so that contact with it, and with a good earth (in this case the tap) makes the human body part of the shock circuit.

The severity of the shock will depend on the effectiveness of the kettle protective conductor system. If the protective system had zero resistance, a 'dead short' would be caused by the fault and the protecting fuse or circuit breaker would open the circuit. The equivalent circuit shown in {Fig 3.8(b)} assumes that the protective conductor has a resistance of twice that of the phase conductor at 0.6 Ω, and will result in a potential difference of 160 V across the victim. The higher the protective circuit resistance, the greater will be the shock voltage, until an open circuit protective system will result in a 240 V shock.

If the protective conductor had no resistance during the short time it took for the circuit to open, the victim would be connected across a zero resistance which would result in no volt drop regardless of the level reached by the fault current, so there could be no shock.

The shock level thus depends entirely on the resistance of the protective system. The lower it can be made, the less severe will be the shocks which may be received.

To sum up this subsection: *Direct contact* is contact with a live system which should be known to be dangerous and *Indirect contact* concerns contact with metalwork which would be expected to be at earth potential, and thus safe. The presence of socket outlets close to sinks and taps is not prohibited by the IEE Wiring Regulations, but could cause danger in some circumstances. It is suggested that special care be taken, including consultation with the Health and Safety Executive in industrial and commercial situations.

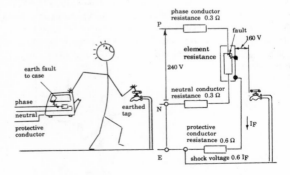

Fig 3.8   Indirect contact Left-side) fault condition
Right-side) equivalent circuit, assuming a fault resistance of zero

3.4.4   ***Protection from contact*** [410, 411, 471-02, 471-03 &471-14]
Four methods of protection are listed in the Regulations.
1   Protection by separated extra-low voltage (SELV)
This voltage is electrically separated from earth and from other sytems, is
provided by a safety source, and is low enough to ensure that contact with it
cannot produce a dangerous shock in people with normal body resistance or
in livestock. The system is uncommon.
2   Protection by protective extra-low voltage (PELV)
The method has the same requirements as SELV but is earthed at one point. Protection
against direct contact may not be required if the equipment is in a building, if the output
voltage level does not exceed 25 V rms or 60 V ripple-free dc in normally dry locations, or 6 V rms ac or 15 V ripple-free dc in all other locations.
3   Protection by functional extra-low voltage (FELV)
This system uses the same safe voltage levels as SELV, but not all the protective measures required for SELV are needed and the system is widely used for
supplies to power tools on construction sites {7.5}. The voltage must not exceed 50 V ac or 120 V dc. The reason for the difference is partly that direct
voltage is not so likely to produce harmful shock effects in the human body as
alternating current, and partly because the stated value of alternating voltage
is r.m.s. and not maximum. As {Fig 3.9} shows, such a voltage rises to a peak
of nearly 71 V, and in some circumstances twice this voltage level may be
present. The allowable 120 V dc must be ripple free. {Figure 3.10} shows
how a 120 V direct voltage with an 80 V peak-to-peak ripple will give a peak
voltage of 160 V. The allowable ripple is such that a 120 V system must

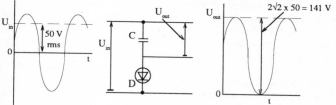

*Fig 3.9   An alternating supply of 50 V may provide 141 V when the
             supply is rectified*

never rise above 140 V or a 60 V system above 70 V. It is interesting to note
that a direct voltage with a superimposed ripple is more likely to cause heart
fibrillation {3.4.2} than one which has a steady voltage.Unlike the SELV sytem,
functional extra low voltage supplies are earthed as a normal installation. Direct contact is prevented by enclosures giving protection to IP2X (which means
that live parts cannot be touched from outside by a human finger — *see* {Ta-
ble 2.4}) or by insulation capable of withstanding 500 V for one minute.

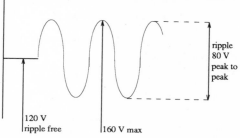

Fig 3.10   Increased peak value of a direct voltage with a ripple

It must be quite impossible for the low voltage levels of the normal installation to appear on the SELV system, and enclosures/insulation used for their separation must be subjected to the same insulation resistance tests as for the higher voltage. Any plugs used in such a circuit must not be interchangeable with those used on the higher voltage system. This will prevent accidentally applying a low voltage to an extra low voltage circuit.

4   *Protection by limitation of discharge energy*

Most electrical systems are capable of providing more than enough energy to cause death by electric shock. In some cases, there is too little energy to cause severe damage. For example, most electricians will be conversant with the battery-operated insulation resistance tester. Although the device operates at a lethal voltage (seldom less than 500 V dc) the battery is not usually capable of providing enough energy to give a fatal shock. In addition, the internal resistance of the instrument is high enough to cause a volt drop which reduces supply voltage to a safe value before the current reaches a dangerous level. This does not mean that the device is safe: it can still give shocks which may result in dangerous falls or other physical or mental problems.

The electric cattle fence is a very good example of a system with limited energy. The system is capable of providing a painful shock to livestock, but not of killing the animals, which are much more susceptible to the effects of shock than humans.

3.4.5   *Direct contact protection* ~ [412, 471-04 to 471-07]

The methods of preventing direct contact are mainly concerned with making sure that people cannot touch live conductors. These methods include:

1.   the insulation of live parts — this is the standard method. The insulated conductors should be further protected by sheathing, conduit, *etc.*

2.   the provision of barriers, obstacles or enclosures to prevent touching (IP2X). Where surfaces are horizontal and accessible, IP4X protection (solid objects wider than 1 mm are excluded — *see* {Table 2.4}), applies

3.   placing out of reach or the provision of obstacles to prevent people from reaching live parts

4.   the provision of residual current devices (RCDs) provides supplementary protection {5.9} but only when contact is from a live part to an earthed part.

3.4.6   *Indirect contact protection* ~ [413, 471-08 to 471-13]

There are two methods of providing protection from shock after contact with a conductor which would not normally be live:

1.   making sure that when a fault occurs and makes the parts live, it results in the supply being cut off within a safe time. In practice, this involves limitation of earth fault loop impedance, a subject dealt with in greater detail in {5.3}.

2.   cutting off the supply before a fatal shock can be received using a residual current device {5.9}.

It is important to appreciate that in some cases a dangerous voltage may be maintained if an uninterruptible power supply (UPS) or a standby generator with automatic starting is in use.

3.4.7   *Protection for users of equipment outdoors* ~ [412-06 and 471-16]

Because of the reduced resistance to earth which is likely for people outdoors, special requirements apply to portable equipments used by them. Experience shows that many accidents have occurred whilst people are using lawn mowers, hedge trimmers, *etc*. There must be a suitable plug (a weather-proof type if installed outdoors) installed in a position which will not result in the need for flexible cords of excessive length. The plug must be protected by an RCD with a rating no greater than 30 mA. Where equipment is fed from a transformer (*eg* the centre-tapped 110 V system on construction sites), the transformer primary must be fed through a 30 mA rated RCD.

These requirements do not apply to extra-low voltage sockets installed for special purposes, such as exterior low voltage garden lighting (*see* {7.14.2}).

3.5   High temperature protection

3.5.1   *Introduction* ~ [13-18, 421-01-01, and554-07]

The Regulations are intended to prevent both fires and burns which arise from electrical causes. Equipment must be selected and installed with the prevention of fire and burns fully considered. Section 421, added in the 1994 amendments, require that persons, equipment and materials adjacent to electrical equipment must be protected from fire, burns and effects limiting the safe functioning of equipment. Three categories of thermal hazard are associated with an electrical installation.

1.   ignition arising directly from the installation,
2.   the spread of fire along cable runs or through trunking where proper fire stops have not been provided, and
3.   burns from electrical equipment.

The heat from direct sunlight will add significantly to the temperature of cables, and 20°C must be added to the ambient temperature when derating a cable subject to direct sunlight, unless it is permanently shaded in a way which does not reduce ventilation. Account must also be taken of the effect of the ultra-violet content of sunlight on the sheath and insulation of some types of cable.

Some types of electrical equipment are intended to become hot in normal service, and special attention is needed in these cases. For example, electric surface heating systems must comply fully with all three parts of BS 6351. Part 1 concerns the manufacture and design of the equipment itself, Part 2 with the design of the system in which it is used, and Part 3 its installation, maintenance and testing.

3.5.2   *Fire protection* ~ [422, 424 and 527]

Where an electrical installation or a piece of equipment which is part of it is, under normal circumstances, likely to become hot enough to set fire to material close to it, it must be enclosed in heat and fire resistant material which will prevent danger. Because of the complexity of the subject, the Regulations give no specific guidance concerning materials or clearance dimensions. It is left to the designer to take account of the circumstances arising in a particular situation. When fixed equipment is chosen by the installation user or by some other party than the designer or the installer, the latter are still responsible for ensuring that the installation requirements of the manufacturers are met.

The same general principle applies in cases where an equipment may emit arcs or hot particles under fault conditions, including arc welding sets. Whilst it may be impossible in every case to prevent the outbreak of fire, attention must be paid to the means of preventing its spread.

For example, any equipment which contains more than 25 litres of flammable liquid, must be so positioned and installed that burning liquid cannot escape the vicinity of the equipment and thus spread fire. {Figure 3.11} indicates the enclosure needed for such a piece of equipment, for example an oil-filled transformer.

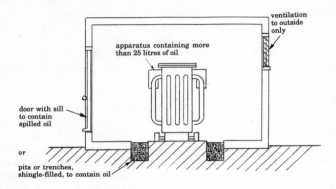

Fig 3.11   Precautions for equipment which contains flammable liquid

Perhaps a word is needed here concerning the use of the word 'flammable'. It means something which can catch fire and burn. We still see the word inflammable in everyday use, with the same meaning as flammable. This is very confusing, because the prefix 'in' may be taken as meaning 'not', giving exactly the opposite meaning. 'Inflammable' should NEVER be used, 'nonflammable' being the correct term for something which cannot catch fire.

Under some conditions, especially where a heavy current is broken, the current may continue to flow through the air in the form of an arc. This is more likely if the air concerned is polluted with dust, smoke, etc. The arc will be extremely hot and is likely to cause burns to both equipments and to people; metal melted by the arc may be emitted from it in the form of extremely hot particles which will themselves cause fires and burns unless precautions are taken. Special materials which are capable of withstanding such arc damage are available and must be used to screen and protect surroundings from the arc.

Some types of electrical equipment, notably spotlights and halogen heaters, project considerable radiant heat. The installer must consider the materials which are subject to this heat to ensure that fire will not occur. Enclosures of electrical equipment must be suitably heat-resisting.

Additions to an installation or changes in the use of the area it serves may give rise to fire risks. Examples are the addition of thermal insulation, the installation of additional cables in conduit or trunking, dust or dirt which restricts ventilation openings or forms an explosive mixture with air, changing lamps for others of higher rating, missing covers on joint boxes and other enclosures so that vermin may attack cables, and so on.

3.5.3   *Protection from burns* ~ [423, 424 and 512-02-01]

The Regulations provide a Table showing the maximum allowable temperatures of surfaces which could be touched and thus cause burns. The allowable temperature depends on whether the surface is metallic or nonmetallic, and on the likely contact between the hand and the surface. Details follow in {Table 3.1}.

Table 3.1 Allowable surface temperatures for accessible parts
(taken from [Table 42A] of BS 7671: 1992)

Part	Surface material	Max. temp (°C)
Hand held	metallic	55
	non-metallic	65
May be touched but not held	metallic	70
	non-metallic	80
Need not be touched in normal use	metallic	80
	non-metallic	90

Other measures intended to prevent water and hot air systems causing burns are contained in Section 424, which was added in the 1994 amendments. They include the requirement that the elements of forced air heaters cannot be switched on until the rate of air flow across them is sufficient to ensure that the air emitted is not too hot, and that water heaters and steam raisers are provided with non self-resetting controls where appropriate. The suitability for connection of high temperature cables must be established with the manufacturer before cables running at more than 70°C are connected.

3.6   Overcurrent protection

3.6.1   *Introduction* ~ [130-03, 431-01, 432-01, 531-01 and 531-2,533-01-01]

'Overcurrent' means what it says — a greater level of current than the materials in use will tolerate for a long period of time. The term can be divided into two types of excess current.

1   Overload currents

These are currents higher than those intended to be present in the system. If such currents persist they will result in an increase in conductor temperature, and hence a rise in insulation temperature. High conductor temperatures are of little consequence except that the resistance of the conductor will be increased leading to greater levels of voltage drop.

Insulation cannot tolerate high temperatures since they will lead to deterioration and eventually failure. The most common insulation material is p.v.c. If it becomes too hot it softens, allowing conductors which press against it (and this will happen in all cases where a conductor is bent) to migrate through it so that they come close to, or even move beyond, the insulation surface. For this reason, p.v.c. insulation should not normally run at temperatures higher than 70°C, whereas under overload conditions it may have allowable temperature up to 115°C for short periods.

2   Short circuit currents

These currents will only occur under fault conditions, and may be very high indeed. As we shall shortly show (*see* {3.6.3 and 3.6.4}) such currents will open the protective devices very quickly. These currents will not flow for long periods, so that under such short-term circumstances the temperature of p.v.c. insulation may be allowed to rise to 160°C.

The clearance time of the protective device is governed by the adiabatic equation which is considered more fully in {3.7.3}.

3.6.2   *Overload* ~ [432-02, 432-03, 433, 473-01, 473-02-01, 473-02-02, 533-02, 533-03]

Overload currents occur in circuits which have no faults but are carrying a higher current than the design value due to overloaded machines, an error in the assessment of diversity, and so on. When a conductor system carries more current than its design value, there is a danger of the conductors, and hence the insulation, reaching temperatures which will reduce the useful life of the system.

The devices used to detect such overloads, and to break the circuit for protection against them, fall into three main categories:

1. Semi-enclosed (rewirable) fuses to BS 3036 and cartridge fuses for use in plugs to BS 1362.
2. High breaking capacity (HBC) fuses to BS 88 and BS 1361. These fuses are still often known as high rupturing capacity (HRC) types.
3. Circuit breakers, miniature and moulded case types to BS EN 60898.

Examination of the characteristics of these devices {Figs 3.13 to 3.19} indicates that they are not the 'instant protectors' they are widely assumed to be. For example, an overloaded 30 A semi-enclosed fuse takes about 100 s to 'blow' when carrying twice its rated current. If it carries 450 A in the event of a fault (fifteen times rated current), it takes about 0.1 s to operate, or five complete cycles of a 50 Hz supply.

HBC fuses are faster in operation, but BS 88 Part 2 specifies that a fuse rated at 63 A or less must NOT operate within one hour when carrying a current 20% greater than its rating. For higher rated fuses, operation must not be within four hours at the same percentage overload. The latter are only required to operate within four hours when carrying 60% more current than their rated value.

Circuit breakers are slower in operation than is generally believed. For example, BS EN 60898 only requires a 30 A miniature circuit breaker to operate within one hour when carrying a current of 40 A. At very high currents operation is described by the BS as 'instantaneous' which is actually within 0.01 seconds.

All protective devices, then, will carry overload currents for significant times without opening. The designer must take this fact into account in his calculations. The circuit must be designed to prevent, as far as possible, the presence of comparatively small overloads of long duration.

The overload provisions of the Regulations are met if the setting of the device:

1. exceeds the circuit design current
2. does NOT exceed the rating of the smallest cable protected

In addition, the current for operation must not be greater than 1.45 times the rating of the smallest cable protected.

The overload protection can be placed anywhere along the run of a cable provided there are no branches, or must be at the point of cable size reduction where this occurs. There must be NO protection in the secondary circuit of a current transformer, or other situation where operation of the protective device would result in greater danger than that caused by the overload. Fuses and circuit breakers controlling a small installation are commonly grouped in a consumer's unit at the mains position. Backless types are still available, and they must be fitted with a non-combustible back on installation.

There are some circuits which have widely varying loads, and it would be unfortunate if the protection operated due to a severe but short-lived overload. In such cases, the heating effect of the currents must be taken into account so that the overload setting is based on the thermal loading.

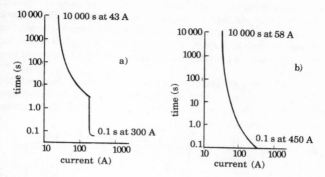

Fig 3.12   *Time/current characteristics a) 30 A miniature circuit breaker*
            type 3   b)   30 A semi-enclosed fuse

3.6.3   *Fuses* ~ [533-01-01 to 533-01-04]

Fuses operate because the fuse element is the 'weak link' in the circuit, so that overcurrent will melt it and break the circuit. The time taken for the fuse link to break the circuit (to 'blow') varies depending on the type of fuse and on the characteristic of the device. The time/current characteristic of a typical fuse is shown in {Fig 3.12(a)}. Curves for other types and ratings of fuses are shown in {Figs 3.13 to 3.15}. The figures are adapted from Appendix 3 of the 16th edition of the IEE Wiring Regulations.

Where the current carried is very much greater than the rated value (which is usually associated with a fault rather than with an overload) operation is usually very fast. For small overloads, where the current is not much larger than the rated value, operation may take a very long time, as indicated.

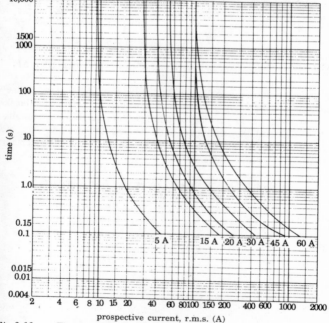

Fig 3.13   *Time/current characteristics of semi-enclosed fuses to BS 3036*

29

A graph with linear axes would need to be very large indeed if the high current/short time and the low current/long time ends of the characteristic were to be used to read the time to operate for a given current. The problem is removed by using logarithmic scales, which open out the low current and short time portions of the scales, and compress the high current and long time portions.

This means that the space between two major lines on the axes of the graph represents a change of ten times that represented by the two adjacent lines. In other words, a very much increased range of values can be accommodated on a graph of a given size.

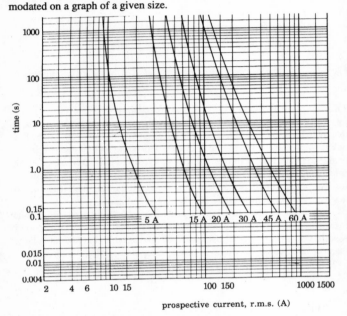

Fig 3.14   Time/current characteristics of cartridge fuses to BS 1361

Rewirable (semi-enclosed) fuses to BS 3036 may still be used, but as they can easily have the wrong fuse element (fuse wire) fitted and have low breaking capacity {3.7.2} they are not recommended for other than small installations. Where used, they are subject to the derating requirements which are explained in {4.3.8}. The diameter of copper wires for use as elements in such fuses is shown in {Table 3.2}.

Table 3.2   Sizes of tinned copper wire fuse elements
(from [Table 53A] of BS 7671: 1992)

Fuse element rating (A)	Wire diameter (mm)
3	0.15
5	0.20
10	0.35
15	0.50
20	0.60
25	0.75
30	0.85
45	1.25
60	1.53

All fuses must be clearly labelled with the fuse rating to make replacement with the wrong fuse as unlikely as possible. It must not be hazardous to make or break a circuit by insertion or removal of a fuse.

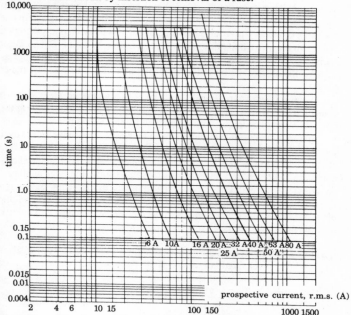

Fig 3.15   Time/current characteristics of cartridge fuses to BS 88 Part 2

3.6.4   *Circuit breakers* ~ [533-01-05 and 533-01-06]

Circuit breakers operate using one or both of two principles. They are:

1.   *Thermal operation* relies on the extra heat produced by the high current warming a bimetal strip, which bends to trip the operating contacts.

2.   *Magnetic operation* is due to the magnetic field set up by a coil carrying the current, which attracts an iron part to trip the breaker when the current becomes large enough.

Thermal operation is slow, so it is not suitable for the speedy disconnection required to clear fault currents. However, it is ideal for operation in the event of small but prolonged overload currents. Magnetic operation can be very fast and so it is used for breaking fault currents; in many cases, both thermal and magnetic operation are combined to make the circuit breaker more suitable for both overload and fault protection. It must be remembered that the mechanical operation of opening the contacts takes a definite minimum time, typically 20 ms, so there can never be the possibility of truly instantaneous operation.

All circuit breakers must have an indication of their current rating. Miniature circuit breakers have fixed ratings but moulded case types can be adjusted. Such adjustment must require the use of a key or a tool so that the rating is unlikely to be altered except by a skilled or instructed person.

There are many types and ratings of moulded case circuit breakers, and if they are used, reference should be made to supplier's literature for their characteristics. Miniature circuit breakers are manufactured in fixed ratings from 5 A to 100 A for some types, and in six types, type B giving the closest protection. Operating characteristics for some of the more commonly used ratings of types 1, 3, B and D are shown in {Figs 3.16 to 3.19}. The characteristics of Type C circuit breakers are very similar to those of Type 3.

BS3871, which specified the miniature circuit breakers Types 1 to 4 was withdrawn in 1994 and has been replaced with BS EN 60898:1991 (EN stands for "European norm"), although it is possible that circuit breakers to the old standard will still be on sale for five years from its withdrawal. In due course, it is intended that only types B, C and D will be available, although it will be many years before the older types cease to be used. Short circuit ratings for the newer types will be a minimum of 3 kA and may be as high as 25 kA - the older types had short circuit ratings which were rarely higher than 9 kA.

It can be seen from {Figs 3.16 to 3.19} that all circuit breakers have characteristics with a vertical section, meaning that at a certain current the possible range of operating times is quite wide. This means that there will be no difference in allowable minimum earth fault loop impedance values over the range of times concerned, which are shown for the six types of miniature circuit breaker in {Table 3.3}. Thus, there will be no difference in the allowable earth-fault loop impedance values for socket outlet and fixed appliance circuits (*see* {Table 5.2}). Over this range of times, the operating current is a fixed multiple of the rated current. For example, a Type 2 MCB has a multiple of x 7 (from {Table 3.3}) so a 30 A rated device of this type will operate over the time range of 0.04 s to 8 s at a current of 7 x 30 A = 210 A.

Table 3.3 Operating time ranges and current multiples for MCBs over fixed current section of characteristic
(from [figs 4 to 8, Appendix 3] of BS 7671: 1992)

MCB Type	Range of operating times (s)	Current multiple of rating
1	0.04 to 20	x 4
2	0.04 to 8	x 7
3	0.04 to 5	x 10
B	0.04 to 13	x 5
C	0.04 to 5	x 10
D	0.05 to 3	x 20

Table 3.4 A comparison of types of protective device.

Semi-enclosed fuses	HBC fuses	Miniature circuit breakers
very low initial cost	medium initial cost	high initial cost
low replacement cost	medium replacement cost	zero replacement cost
low breaking capacity	very high breaking capacity	medium breaking capacity

Table 3.5 Comparison of miniature circuit breaker types

Type	Will NOT trip in 100 ms at rating	Will trip in 100ms at rating	Typical application
1	2.7 x	4 x	low inrush currents (domestic installations)
2	4 x	7 x	general purpose use
3	7 x	10 x	high inrush currents (motor circuits)
B	3 x	5 x	general purpose use (close protection)
C	5 x	10 x	commercial and industrial applications with fluorescent fittings
D	10 x	50 x	applications where high in-rush currents are likely (transformers, welding machines)

{Table 3.4} shows a comparison of the three main types of protective device in terms of cost, whilst {Table 3.5} compares the available types of MCB.

3.6.5   *Protecting conductors* ~[433-01 and 433-02, 473-02, 473-03]
The prime function of overload protection is to safeguard conductors and cables from becoming too hot. Thus the fuse or circuit breaker rating must be no greater than that of the smallest cable protected. Reference to the time/current characteristics of protective devices {Figs 3.13 to 3.19} shows that a significantly greater current than the rated value is needed to ensure operation.

Thus, the current at which the protective device operates must never be greater than 1.45 times the rating of the smallest cable protected. For example, consider a cable system rated at 30 A and protected by a miniature circuit breaker type 3, also rated at 30 A. Reference to {Fig 3.18} shows that a prolonged overload of about 38 A will open the breaker after about 10^4 seconds (about two and a half hours!). The ratio of operating current over rated current is thus 38/30 or 1.27, significantly lower than the maximum of 1.45. All circuit breakers and HBC fuses listed in {3.6.2 sections 2 and 3} will comply with the Regulations as long as their rating does not exceed that of the smallest cable protected.

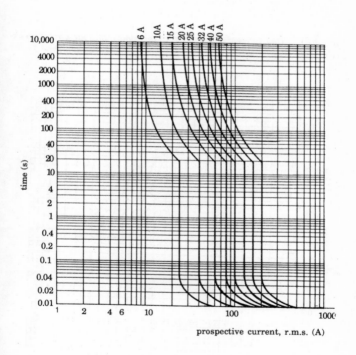

prospective current, r.m.s. (A)

Fig 3.16   *Time/current characteristics for some miniature circuit breakers Type 1*

33

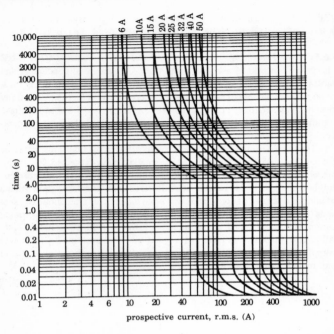

Fig 3.17   Time/current characteristics for some miniature circuit breakers Type 3. Type C MCBs have very similar characteristics to Type 3

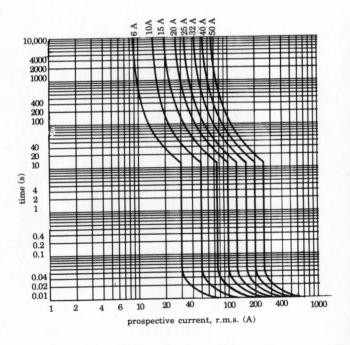

Fig 3.18   Time/current characteristics for some miniature circuit breakers Type B

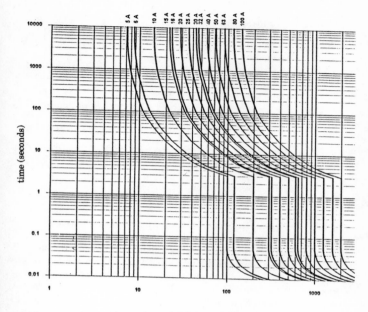

prospective current, rms (amperes)

Fig 3.19   Time/current characteristics for some miniature circuit breakers Type D

Semi-enclosed (rewirable) fuses do not operate so closely to their ratings as do circuit breakers and HBC fuses. For example, the time/current characteristics of {Fig 3.13} show that about 53 A is needed to ensure the operation of a 30 A fuse after 10,000 s, giving a ratio of 53/30 or 1.77. For rewirable fuses, the Regulations require that the fuse current rating must not exceed 0.725 times the rating of the smallest cable protected. Considering the 30 A cable protected by the 30 A miniature circuit breaker above, if a rewirable fuse replaced the circuit breaker, its rating must not be greater than 0.725 x 30 or 21.8 A.

Since overload protection is related to the current-carrying capacity of the cables protected, it follows that any reduction in this capacity requires overload protection at the point of reduction. Reduced current-carrying capacity may be due to any one or more of:

1. a reduction in the cross-sectional area of the cable
2. a different type of cable
3. the cable differently installed so that its ability to lose heat is reduced
4. a change in the ambient temperature to which the cable is subjected
5. the cable is grouped with others.

{Figure 3.20} shows part of a system to indicate how protection could be applied to conductors with reduced current carrying capacity.

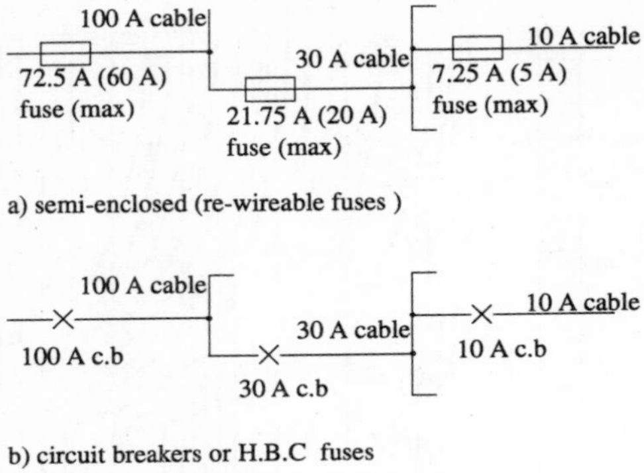

a) semi-enclosed (re-wireable fuses)

b) circuit breakers or H.B.C fuses

Fig 3.20   Position and rating of devices for overload protection

In fact, the calculated fuse sizes for {Fig 3.20a)} of 72.5 A, 21.75 A and 7.25 A are not available, so the next lowest sizes of 60 A, 20 A and 5 A respectively must be used. It would be unwise to replace circuit breakers with semi-enclosed fuses because difficulties are likely to arise. For example, the 5 A fuse used as the nearest practical size below 7.25 A is shown in {Fig 3.13} to operate in 100 s when carrying a current of 10 A. Thus, if the final circuit is actually carrying 10 A, replacing a 10 A circuit breaker with a 5 A fuse will result in the opening of the circuit. The temptation may be to use the next semi-enclosed fuse size of 15 A, but that fuse takes nearly seven minutes to operate at a current of 30 A. Clearly, the cable could well be damaged by excessive temperature if overloaded.

The device protecting against overload may be positioned on the load side of (downstream from) the point of reduction, provided that the unprotected cable length does not exceed 3 m, that fault current is unlikely, and that the cable is not in a position that is hazardous from the point of view of ignition of its surroundings. This Regulation is useful when designing switchboards, where a short length of cable protected by conduit or trunking feeds a low-current switch fuse from a high current fuse as in {Fig 6.2}.

All phase conductors must be protected, but attention must be paid to the need to break at the same time all three line conductors to a three-phase motor in the event of a fault on one phase, to prevent the motor from being damaged by 'single-phasing'. Normally the neutral of a three phase system should not be broken, because this could lead to high voltages if the load is unbalanced. Where the neutral is of reduced size, overload protection of the neutral conductor may be necessary, but then a circuit breaker must be used so that the phases are also broken.

3.7   Protection from faults
3.7.1   *Introduction*   ~   [432-02, 432-04, 434]
The overload currents considered in the previous section are never likely to be more than two to three times the normal rated current. Fault currents, on the other hand, can well be several hundreds, or even several thousands of times normal. In the event of a short circuit or an earth fault causing such current, the circuit must be broken before the cables are damaged by high temperatures or by electro-mechanical stresses. The latter stresses will be due to the force on a current carrying conductor which is subject to the mag-

netic field set up by adjacent conductors. This force is proportional to the current, and to the magnetic field strength. Since the field strength also depends on the current, force is proportional to the square of the current. If the current is one thousand times normal, force will be one million times greater than usual! Fault protection must not only be able to break such currents, but to do so before damage results. Abrasion of cable insulation by movement is usually prevented by normal fixings or by being enclosed in conduit or trunking. Support must be provided to cables in busbar chambers.

3.7.2   *Prospective short-circuit current (PSC)* ~ [313-01-01, 432-02, 434-02]

The current which is likely to flow in a circuit if line and neutral cables are short circuited is called the prospective short circuit current (PSC). It is the largest current which can flow in the system, and protective devices must be capable of breaking it safely. The breaking capacity of a fuse or of a circuit breaker is one of the factors which need to be considered in its selection. Consumer units to BS EN 60439-4 and BS 88 (HBC) fuses are capable of breaking any probable prospective short-circuit current, but before using other equipments the installer must make sure that their breaking capacity exceeds the PSC at the point at which they are to be installed.

The effective breaking capacity of overcurrent devices varies widely with their construction. Semi-enclosed fuses are capable of breaking currents of 1 kA to 4 kA depending on their type, whilst cartridge fuses to BS 1361 will safely break at 16.5 kA for type I or 33 kA for type II. BS 88 fuses are capable of breaking any possible short-circuit current. Miniature circuit breakers to BS EN60898 have their rated breaking capacity marked on their cases in amperes (not kA) although above 10000 A the MCB may be damaged and lower breaking currents (75% for 10000 A and 50% above that level) must be used for design purposes.

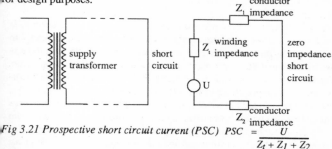

Fig 3.21 Prospective short circuit current (PSC) $PSC = \dfrac{U}{Z_t + Z_1 + Z_2}$

Prospective short circuit current is driven by the e.m.f. of he secondary winding of the supply transformer through an impedance made up of the secondary winding and the cables from the transformer to the fault {Fig 3.21}. The impedance of the cables will depend on their size and length, so the PSC value will vary throughout the installation, becoming smaller as the distance from the intake position increases. [313-01-01] requires the PSC to be 'assessed' by 'calculation, measurement, enquiry or inspection'. In practice, this can be difficult because it depends to some extent on impedances which are not only outside the installation in the supply system, but are also live. If the impedance of the supply system can be found, a straightforward calculation using the formula of {Fig 3.21} can be used, but this is seldom the case. An alternative is to ask the local Electricity Company. The problem here is that they are likely to protect themselves by giving a figure which is usually at least 16 kA in excess of the true value. The problem with using this figure is that the higher the breaking capacity of fuses and circuit breakers are (and this must never be less than the PSC for the point at which they are installed), the higher will be their cost. {Table 3.6} gives a method of arriving at PSC if the type and length of the service cable is known.

Table 3.6	Estimation of PSC at the intake position	
Length of supply cable (m)	*PSC (kA) up to 25 mm² Al, 16 mm² Cu supply cable*	*PSC (kA) over 35 mm² Al 25 mm² Cu supply cable*
5	10.0	12.0
10	7.8	9.3
15	6.0	7.4
20	4.9	6.2
25	4.1	5.3
30	3.5	4.6
40	2.7	3.6
50	2	3.0

The table is not applicable in London, where the density of the distribution system means that higher values may apply. In this case it will be necessary to consult London Electricity.

There are two methods for measuring the value of PSC, but these can only be used when the supply has already been connected. By then, the fuses and circuit breakers will already be installed.

The first method is to measure the impedance of the supply by determining its voltage regulation, that is, the amount by which the voltage falls with an increase in current. For example, consider an installation with a no-load terminal voltage of 240 V. If, when a current of 40 A flows, the voltage falls to 238 V, the volt drop will be due to the impedance of the supply.

$$\text{Thus } Z_S = \frac{\text{system volt drop}}{\text{current}} = \frac{240 - 238}{40} \, \Omega = \frac{2}{40} \, \Omega = 0.05 \, \Omega$$

$$\text{Then PSC} = \frac{U_O}{Z_S} = \frac{240}{0.05} \, A = 4800 \, A \text{ or } 4.8 \, kA$$

A second measurement method is to use a loop impedance tester {*see* 5.3 and 8.6.2} connected to phase and neutral (instead of phase and earth) to measure supply impedance. This can then be used with the supply voltage as above to calculate PSC. Some manufacturers modify their earth-loop testers so that this connection is made by selecting 'PSC' with a switch. The instrument measures supply voltage, and calculates, then displays, PSC.

A possible difficulty in measuring PSC, and thus being able to use fuses or circuit breakers with a lower breaking capacity than that suggested by the Supply Company, is that the supply may be reinforced. More load may result in extra or different transformers and cables being installed, which may reduce supply impedance and increase PSC.

3.7.3 *Operating time* ~ [434-01, 434-03]

Not only must the short-circuit protection system open the circuit to cut off a fault, but it must do so quickly enough to prevent both a damaging rise in the conductor insulation temperature and mechanical damage due to cable movement under the influence of electro-mechanical force. The time taken for the operation of fuses and circuit breakers of various types and ratings is shown in {Figs 3.13 to 3.19}. When the prospective short circuit current (PSC) for the point at which the protection is installed is less than its breaking capacity there will be no problem.

The time taken to clear the fault must, as explained, be limited to a value given by the adiabatic equation.

The position is complicated because the rise in conductor temperature results in an increase in resistance which leads to an increased loss of energy and increased heating ($W = I^2Rt$). The Regulations make use of the adiabatic equation which assumes that all the energy dissipated in the conductor remains within it in the form of heat, because the faulty circuit is opened so quickly. The equation is:

$$t = \frac{k^2S^2}{I^2}$$

where   t   = the time for fault current to raise conductor temperature to the highest permissible level
         k   = a factor which varies with the type of cable
         S   = the cross-sectional area of the conductor (mm^2)
         I   = the fault current value (A) - this will be the PSC

Some cable temperatures and values of k for common cables are given in {Table 3.7}.

Table 3.7   Cable temperatures and k values (copper cable)
(from [Table 43A] of BS 7671: 1992)

Insulation material	Assumed initial temperature (^oC)	Limiting final temperature (^oC)	k
p.v.c.	70	160	115
85°C p.v.c.	85	160	104
mineral, exposed to touch or p.v.c. covered mineral, not exposed to touch	70	160	115
	105	250	135

As an example, consider a 10 mm^2 cable with p.v.c. insulation protected by a 40 A fuse to BS 88 Part 2 in an installation where the loop impedance between lines at the point where the fuse is installed is 0.12 Ω. If the supply is 415 V three phase, the prospective short circuit current (PSC) will be:

$$I = \frac{U_L}{Z} \, A = \frac{415}{0.12} \, A = 3.46 \, kA$$

From {Table 3.7},   k = 115

$$t = \frac{k^2S^2}{I^2} = \frac{115^2 \times 10^2}{3460^2} \, s = 0.110 \, s$$

{Figure 3.15} shows that a 40 A fuse to BS 88 Part 2 will operate in 0.1 s when carrying a current of 400 A. Since the calculated PSC at 3460 A is much greater than 400 A, the fuse will almost certainly clear the fault in a good deal less than 0.1 s. As this time is less than that calculated by using the adiabatic equation (0.11 s) the cable will be unharmed in the event of a short circuit fault.

It is important to appreciate that the adiabatic equation applies to all cables, regardless of size. Provided that a protective device on the load side of a circuit has a breaking capacity equal to or larger than the PSC of the circuit then that circuit complies with the PSC requirements of the Regulations (*see* {Fig 3.22} and *see also* the note in {7.15.1} concerning the use of dual rated fuses for motor protection).

3.7.4   *Conductors of reduced current-carrying capacity* ~ [473-02-02]

Short circuit protection must be positioned at every point where a reduction in cable-current carrying capacity occurs, as for overload protection {Fig 3.20}. However, if short circuit protection on the supply side of the point of reduction (for example, at the incoming mains position) has a characteristic that protects the reduced conductors, no further protection is necessary {Fig 3.22}.

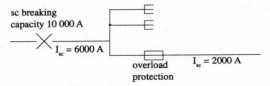

Fig 3.22   Short circuit protective device protecting a circuit of reduced cross-sectional area

Even if not protected by a suitable device on the supply side, short circuit protection may be positioned on the load side of the reduction in rating if the conductors do not exceed 3 m in length, and are protected by trunking and conduit, and are not close to flammable materials. This reduction is particularly useful when connecting switchgear {Fig 3.23}. It should be noted that the 'tails' provided to connect to the supply system should always be of sufficient cross-sectional area to carry the expected maximum demand, and should never be smaller than 25 mm^2 for live conductors and 16 mm^2 for the main earthing conductor.

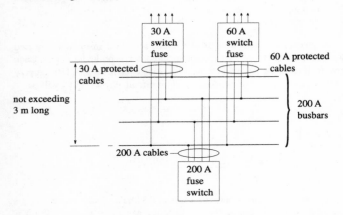

Fig 3.23   Short-circuit protection not required for short switchgear connections

3.7.5   *Back-up protection* ~ [435]

There are times when the overload protection has insufficient breaking capacity safely to interrupt the prospective short circuit current at the point of the installation where it is situated. An instance would be where a large number of low-rated miniature circuit breakers each with a breaking capacity of 3 kA are fed by a large cable. In the event of a short circuit which gives a current of 8 kA, there is a good chance that the miniature circuit breaker concerned will be unable to break the fault.

Perhaps the fault current may continue to flow in the form of an arc across the opened circuit breaker contacts, causing a very high temperature and the danger of fire. Of course, if this happened the circuit breaker would be destroyed. The normal method of protection is to 'back up' the circuit with a protective device which has the necessary breaking capacity. For example, the group of miniature circuit breakers mentioned above could be backed up by an HBC fuse as shown in {Fig 3.24}.

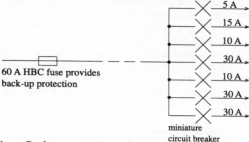

60 A HBC fuse provides
back-up protection

miniature
circuit breaker

Fig 3.24   Back-up protection

It will be appreciated that when a protective device operates it does not do so instantaneously, and fault current will flow through it to the circuit it seeks to protect. The time for which such a current flows is a critical factor in the damage that may be done to the system before the fault clears. Damage applies to cables, switchgear, protective devices, *etc*. if the fault is not cleared quickly enough. This damage will be the result of the release of the energy of the fault current, and the system designer will aim to minimise it by calculation and by consulting manufacturer's data of energy let-through.

3.7.6   *Insulation monitoring* ~ [531-06]

In some special applications, usually IT systems, the insulation effectiveness of an installation is monitored continuously by a device which measures leakage current. If this current exceeds a preset limit, the controlling circuit breaker opens to make the circuit safe. The setting of such a monitor must be arranged so that it can only be altered by an authorised person, and to that end it must be possible to alter the setting only with a special tool.

3.8   Short circuit and overload protection

3.8.1   *Combined protection* ~ [13-7, 432-2, 435]

In many practical applications, most types of fuses and circuit breakers are suitable for both overload and short circuit protection. Care must be taken, however, to ensure that the forms of protection are chosen so that they are properly coordinated to prevent problems related to excessive let-through of energy or to lack of discrimination {3.8.6}.

3.8.2   *Current limited by supply characteristic* ~ [436]

If a supply has a high impedance, the maximum current it can provide could be less than the current carrying capacity of the cables in the installation. In such a case, no overload or short circuit protection is required.

   This situation is very unlikely with a supply taken from an Electricity Company, but could well apply to a private generating plant.

3.8.3   *Protection omitted* ~ [473-01-03, 473-01-04, 473-02-02, 473-02-04]

There are cases where a break in circuit current due to operation of a protective device may cause more danger than the overload or fault. For example, breaking the supply to a lifting electromagnet in a scrap yard will cause it to drop its load suddenly, possibly with dire consequences. If the field circuit of a dc motor is broken, the reduction in field flux may lead to a dangerous increase in speed. A current transformer has many more secondary than primary turns, so dangerously high voltages will occur if the secondary circuit is broken.

In situations like these the installation of an overload alarm will give warning of the faulty circuit, which can be switched off for inspection when it is safe to do so. The possibility of short circuits in such cables will be reduced if they are given extra protection.

Probably the most usual case of omission of protection is at the incoming mains position of a small installation. Here, the supply fuse protects the installation tails and the consumer's unit. The unprotected equipment must, however, comply with the requirements for otherwise unprotected systems listed in {3.7.4}.

3.8.4 Protection of conductors in parallel ~ [433-03, 434-03-02]

The most common application of cables in parallel is in ring final circuits for socket outlets, whose special requirements will be considered in {6.3.2}. Cables may otherwise be connected in parallel provided that they are of exactly the same type and rating, run together throughout their length, have no branches and are expected to share the total circuit current equally.

Overload protection must then be provided for the sum of the current-carrying capacities of the cables. If, for example, two cables with individual current ratings of 13 A are connected in parallel, overload protection must be provided for 26 A.

Account must also be taken of a short circuit which does not affect all the cables: this is made less likely by the requirement that they should run close together. In {Fig 3.25}, for example, the 30 A cable with the short circuit to neutral must be able to carry more than half the short circuit current without damage until the protection opens.

Fault current sharing in these circumstances depends on the inverse of the ratio of the conductor resistances. If, for example, the fault on one cable were to occur close to the connection to the protective device, almost all of the fault current would be carried by the short length from the protection to the fault.

In these circumstances there would be little protection for the faulty cable, and it would be prudent to provide protection with the installation of a suitable RCD.

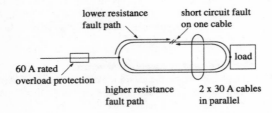

Fig 3.25   Cables in parallel. The lower resistance path will carry the higher fault current

3.8.5   Absence of protection

The protection described in this Chapter applies to conductors, but not necessarily to the equipment fed. This is particularly true where flexible cords are concerned. For example, a short circuit in an appliance fed through a 0.5 mm² flexible cord from a 13 A plug may well result in serious damage to the cord or the equipment before the fuse in the plug can operate.

3.8.6   ***Discrimination***   ~   [531-02, 533-01]

Most installations include a number of protective devices in series, and they must operate correctly relative to each other if healthy circuits are not to be disconnected. Discrimination occurs when the protective device nearest to the fault operates, leaving all other circuits working normally.

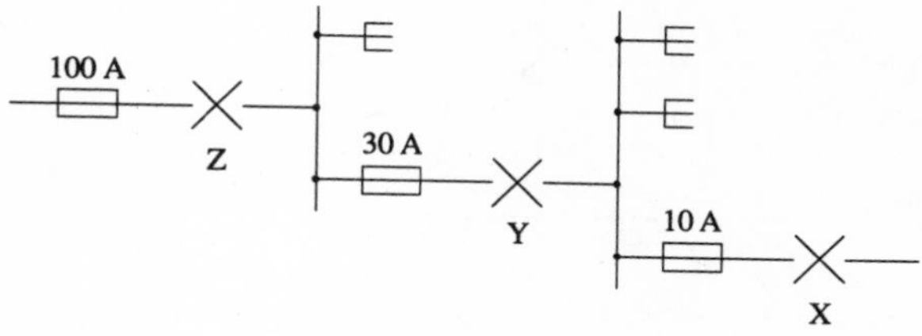

Fig 3.26   *System layout to explain discrimination*

{Figure 3.26} shows an installation with a 100 A main fuse and a 30 A submain fuse feeding a distribution board containing 10 A fuses. If a fault occurs at point Z, the 100 A fuse will operate and the whole installation will be disconnected. If the fault is at X, the 10 A fuse should operate and not the 30 A or 100 A fuses. A fault at Y should operate the 30 A, and not the 100 A fuse. If this happens, the system has discriminated properly.

Lack of discrimination would occur if a fault at X caused operation of the 30 A or 100 A fuses, but not the 10 A fuse. This sounds impossible until we remember the time/current fuse characteristics explained in {3.6.3}. For example, {Fig 3.27} shows the superimposed characteristics of a 5 A semi-enclosed fuse and a 10 A miniature circuit breaker which we shall assume are connected in series.

If a fault current of 50 A flows, the fuse will operate in 0.56 s whilst the circuit breaker would take 24 s to open. Clearly the fuse will operate first and the devices have discriminated. However, if the fault current is 180 A, the circuit breaker will open in 0.016 s, well before the fuse would operate, which would take 0.12 s. In this case, there has been no discrimination.

To ensure discrimination is a very complicated matter, particularly where an installation includes a mixture of types of fuse, or of fuses and circuit breakers. Manufacturers' operating characteristics must be studied to ensure discrimination. As a rule of thumb where fuses or circuit breakers all of the same type are used, there should be a doubling of the rating as each step towards the supply is taken.

When fault current is high enough to result in operation of the protective device within 40 ms (two cycles of a 50 Hz supply), the simple consideration of characteristics as shown in Fig. 3.27 may not always result in correct discrimination and device manufacturers should be consulted.

When RCDs are connected in series, discrimination between them is also important, the rule here being that a trebling in rating applies with each step towards the supply (*see also {8.6.3}*).

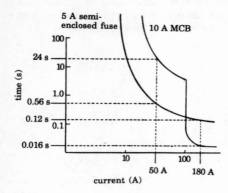

Fig 3.27   To illustrate a lack of discrimination

Cables, conduits and trunking

4.1 Introduction [521-01]

This Chapter is concerned with the selection of wiring cables for use in an electrical installation. It also deals with the methods of supporting such cables, ways in which they can be enclosed to provide additional protection, and how the conductors are identified. All such cables must conform in all respects with the appropriate British Standard.

This Electrician's Guide does *not* deal with cables for use in supply systems, heating cables, or cables for use in the high voltage circuits of signs and special discharge lamps.

4.1.1 *Cable insulation materials*
Rubber

For many years wiring cables were insulated with vulcanised natural rubber (VIR). Much cable of this type is still in service, although it is many years since it was last manufactured. Since the insulation is organic, it is subject to the normal ageing process, becoming hard and brittle. In this condition it will continue to give satisfactory service unless it is disturbed, when the rubber cracks and loses its insulating properties. It is advisable that wiring of this type which is still in service should be replaced by a more modern cable. Synthetic rubber compounds are used widely for insulation and sheathing of cables for flexible and for heavy duty applications. Many variations are possible, with conductor temperature ratings from 60°C to 180°C, as well as resistance to oil, ozone and ultra-violet radiation depending on the formulation.

Paper

Dry paper is an excellent insulator but loses its insulating properties if it becomes wet. Dry paper is hygroscopic, that is, it absorbs moisture from the air. It must be sealed to ensure that there is no contact with the air. Because of this, paper insulated cables are sheathed with impervious materials, lead being the most common. PILC (paper insulated lead covered) is traditionally used for heavy power work. The paper insulation is impregnated with oil or non-draining compound to improve its long-term performance. Cables of this kind need special jointing methods to ensure that the insulation remains sealed. This difficulty, as well as the weight of the cable, has led to the widespread use of p.v.c. and XLPE (thermosetting) insulated cables in place of paper insulated types.

P.V.C

Polyvinyl chloride (p.v.c.) is now the most usual low voltage cable insulation. It is clean to handle and is reasonably resistant to oils and other chemicals. When p.v.c. burns, it emits dense smoke and corrosive hydrogen chloride gas. The physical characteristics of the material change with temperture: when cold it becomes hard and difficult to strip, and so the IEE Regulations specify that it should not be worked at temperatures below 5°C. However a special p.v.c. is available which remains flexible at temperatures down to -20°C.

At high temperatures the material becomes soft so that conductors which are pressing on the insulation (*eg* at bends) will 'migrate' through it, sometimes moving to the edge of the insulation. Because of this property the temperature of general purpose p.v.c. must not be allowed to exceed 70°C, although versions which will operate safely at temperatures up to 85°C are also available. If p.v.c. is exposed to sunlight it may be degraded by ultra-violet radiation. If it is in contact with absorbent materials, the plasticiser may be 'leached out' making the p.v.c. hard and brittle.

LSF (Low smoke and fume)

Materials which have reduced smoke and corrosive gas emissions in fire compared with p.v.c. have been available for some years. They are normally used as sheathing compounds over XLPE or LSF insulation, and can give considerable safety advantages in situations where numbers of people may have to be evacuated in the event of fire.

Thermosetting (XLPE)

Cross-linked polyethylene (XLPE) is a thermosetting compound which has better electrical properties than p.v.c. and is therefore used for medium- and high-voltage applications. It has more resistance to deformation at higher temperatures than p.v.c., which it is gradually replacing. It is also replacing PILC in some applications. Thermosetting insulation may be used safely with conductor temperatures up to 90 °C thus increasing the useful current rating, especially when ambient temperature is high. A LSF (low smoke and fume) type of thermosetting cable is available.

Mineral

Provided that it is kept dry, a mineral insulation such as magnesium oxide is an excellent insulator. Since it is hygroscopic (it absorbs moisture from the air) this insulation is kept sealed within a copper sheath. The resulting cable is totally fireproof and will operate at temperatures of up to 250°C. It is also entirely inorganic and thus non-ageing. These cables have small diameters compared with alternatives, great mechanical strength, are waterproof, resistant to radiation and electro-magnetic pulses, are pliable and corrosion resistant. In cases where the copper sheath may corrode, the cable is used with an overall LSF covering, which reduces the temperature at which the cable may be allowed to operate. Since it is necessary to prevent the ingress of moisture, special seals are used to terminate cables. Special mineral-insulated cables with twisted cores to reduce the effect of electro-magnetic interference are available.

4.2   Cables

4..2.1   *Non-flexible low voltage cables* ~ [521-01-01]
Types of cable currently satisfying the Regulations are shown in {Fig 4.1}.

a)   *Non-armoured pvc-insulated cables*

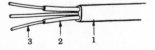

1 - pvc sheath
2 - pvc insulation
3 - copper conduxtor:
solid, stranded or flexible

b)   *Armoured pvc-insulated cables*

1 - pvc sheath
2 - armour-galvanised steel wire
3 - pvc bedding
4 - pvc insulation
5 - copper conductor

Fig 4.1   *Fixed wiring cables*

c) *Split-concentric pvc-insulated cables*

1- pvc oversheath, 2 - Melinex binder, 3 - pvc strings, 4 - neutral conductor: black pvc-covered wires, 5 - earth continuity conductor: bare copper wires,
6 - pvc phase insulation, 7- copper conductors

d) *Rubber-insulated (elastomeric) cables*

1 - textile braided and compounded
2 - 85°C rubber insulation
3 - tinned copper conductor

e) *Impregnated-paper insulated lead sheathed cables*

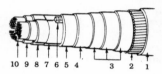

1 - pvc oversheath
2 - galvanised steel wire armour
3 - bedding
4 - sheath: lead or lead alloy
5 - copper woven fabric tape
6 - filler
7 - screen of metal tape intercalated with paper tape
8 - impregnated paper insulation
9 - carbon paper screen
10 - shaped stranded conductor

f) *Armoured cables with thermosetting insulation*

1 - pvc oversheath
2 - galvanised steel wire armour
3 - taped bedding
4 - XLPE insulation
5 - solid aluminium conductor

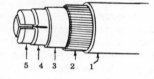

g) *Mineral-insulated cables*

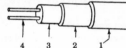

1 - LSF oversheath
2 - copper sheath
3 - magnesium oxide insulation
4 - copper conductors

h) *Consac cables*

1 - extruded pvc or polythene oversheath
2 - thin layer of bitumen containing a corrosion inhibitor
3 - extruded smooth aluminium sheath
4 - paper belt insulation
5 - paper core insulation
6 - solid aluminium conductors

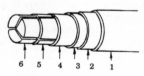

i) *Waveconal cables*

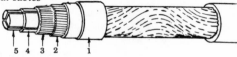

1 - extruded pvc oversheath
2 - aluminium wires
3 - rubber anti-corrosion bedding
4 - XLPE core insulation
5 - solid aluminium conductors

Fig 4.1   Fixed wiring cables (continued)

47

[Table 52B] gives the maximum conductor operating temperatures for the various types of cables. For general purpose p.v.c. this is 70°C. Cables with thermosetting insulation can be operated with conductor temperatures up to 90°C BUT since the accessories to which they are connected may be unable to tolerate such high temperatures, operation at 70°C is much more usual. Other values of interest to the electrician are shown in {Table 3.7}. Minimum cross-sectional areas for cables are shown in {Table 4.1}.

Table 4.1 Minimum permitted cross-sectional areas for cables
(from [Table 52C] of BS 7671: 1992)

type of circuit	conductor material	cross-sectional area (mm²)
power and lighting circuits	copper	1.0
(insulated conductors)	aluminium	16.0
signalling and control circuits	copper	0.5
flexible, more than 7 cores	copper	0.1
bare conductors and busbars	copper	10.0
	aluminium	16.0
bare conductors for signalling and control	copper	4.0

4.2.2 Cables for overhead lines ~ [521-01-03, 522-05, 522-07 & 522-08-01]

Any of the cables listed in the previous subsection are permitted to be used as overhead conductors *provided that they are properly supported*. Normally, of course, the cables used will comply with a British Standard referring particularly to special cables for use as overhead lines. Such cables include those with an internal or external catenary wire, which is usually of steel and is intended to support the weight of the cable over the span concerned.

Since overhead cables are to be installed outdoors, they must be chosen and installed so as to offset the problems of corrosion. Since such cables will usually be in tension, their supports must not damage the cable or its insulation. More information on corrosion is given in {4.2.5} and on the selection and installation of overhead cables will be found in {7.13.1}.

4.2.3 Flexible low voltage cables and cords ~ [521-01-04]

By definition flexible cables have conductors of cross-sectional area 4 mm² or greater, whilst flexible cords are sized at 4 mm² or smaller. Quite clearly, the electrician is nearly always concerned with flexible cords rather than flexible cables.

{Figure 4.2} shows some of the many types of flexible cords which are available.

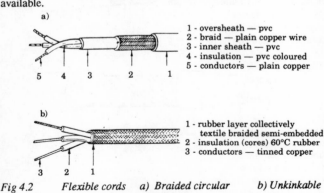

a)
1 - oversheath — pvc
2 - braid — plain copper wire
3 - inner sheath — pvc
4 - insulation — pvc coloured
5 - conductors — plain copper

b)
1 - rubber layer collectively textile braided semi-embedded
2 - insulation (cores) 60°C rubber
3 - conductors — tinned copper

Fig 4.2   *Flexible cords*   *a) Braided circular*   *b) Unkinkable*

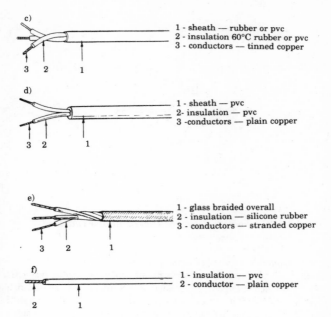

c)
1 - sheath — rubber or pvc
2 - insulation 60°C rubber or pvc
3 - conductors — tinned copper

d)
1 - sheath — pvc
2- insulation — pvc
3 -conductors — plain copper

e)
1 - glass braided overall
2 - insulation — silicone rubber
3 - conductors — stranded copper

f)
1 - insulation — pvc
2 - conductor — plain copper

Fig 4.2   (continued) Flexible cords
c) Circular sheathed
d) Flat twin sheathed
e) Braided circular insulated with glass fibre
f) Single core p.v.c.-insulated non-sheathed

Flexible cables should not normally be used for fixed wiring, but if they are, they must be visible throughout their length. The maximum mass which can be supported by each flexible cord is listed in [Table 4H3A], part of which is shown here as {Table 4.2}.

Table 4.2 Maximum mass supported by twin flexible cord
(from [Table 4H3A] of BS 7671: 1992)

Cross-sectional area (mm²)	Maximum mass to be supported (kg)
0.5	2
0.75	3
1.0	5
1.25	5
1.5	5

The temperature at the cord entry to luminaires is often very high, especially where filament lamps are used. It is important that the cable or flexible cord used for final entry is of a suitable heat resisting type, such as 150°C rubber-insulated and braided. {Fig 4.3} shows a short length of such a cord used to make the final connection to a luminaire.

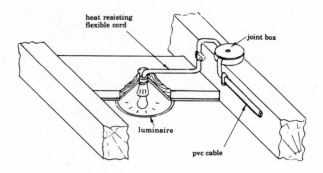

*Fig 4.3   150°C rubber-insulated and braided flexible cord used
for the final connection to a luminaire*

4.2.4   Cables carrying alternating currents ~ [521-02 & 523-05]

Alternating current flowing in a conductor sets up an alternating magnetic field which is much stronger if the conductor is surrounded by an iron-rich material, for example if it is steel wire armoured or if it is installed in a steel conduit. The currents in a twin cable, or in two single core cables feeding a single load, will be the same. They will exert opposite magnetic effects which will almost cancel, so that virtually no magnetic flux is produced if they are both enclosed in the same conduit or armouring. The same is true of three-phase balanced or unbalanced circuits provided that all three (or four, where there is a neutral) cores are within the same steel armouring or steel conduit. An alternating flux in an iron core results in iron losses, which result in power loss appearing as heat in the metal enclosure. It should be remembered that not only will the heat produced by losses raise the temperature of the conductor, but that the energy involved will be paid for by the installation user through his electricity meter. Thus, it is important that all conductors of a circuit are contained within the same cable, or are in the same conduit if they are single-core types (*see* {Fig 4.4}).

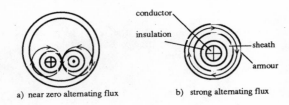

*Fig 4.4   Iron losses in the steel surrounding a cable when it carries
alternating current   a) twin conductors of the same single-
phase circuit — no losses
b) single core conductor— high losses*

A similar problem will occur when single-core conductors enter an enclosure through separate holes in a steel end plate {Fig 4.5}.

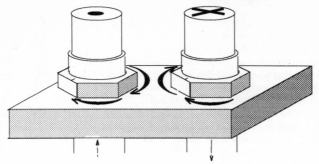

Fig 4.5   Iron losses when single-core cables enter a steel enclosure through separate holes

For this reason, single-core armoured cables should not be used. If the single core cable has a metal sheath which is non-magnetic, less magnetic flux will be produced. However, there will still be induced e.m.f. in the sheath, which can give rise to a circulating current and sheath heating.

If mineral insulated cables are used, or if multi-core cables are used, with all conductors of a particular circuit being in the same cable, no problems will result. The copper sheath is non-magnetic, so the level of magnetic flux will be less than for a steel armoured cable; there will still be enough flux, particularly around a high current cable, to produce a significant induced e.m.f. However, multi-core mineral insulated cables are only made in sizes up to 25 mm^2 and if larger cables are needed they must be single core.

{Figure 4.6(a)} shows the path of circulating currents in the sheaths of such single core cables if both ends are bonded. {Figure 4.6(b)} shows a way of breaking the circuit for circulating currents.

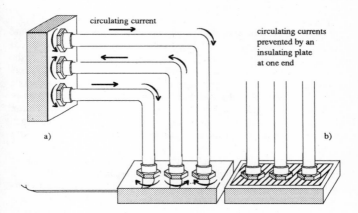

Fig 4.6   Circulating currents in the metal sheaths of single core cables
a) bonded at both ends
b) circulating currents prevented by single point bonding

[523-05-01] calls for all single core cable sheaths to be bonded at both ends unless they have conductors of 70 mm^2 or greater. In that case they can be single point bonded if they have an insulating outer sheath, provided that:

i) e.m.f. values no greater than 25 V to earth are involved, and
ii) the circulating current causes no corrosion, and
iii) there is no danger under fault conditions.

The last requirement is necessary because fault currents will be many times greater than normal load currents. This will result in correspondingly larger values of alternating magnetic flux and of induced e.m.f.

4.2.5   *Corrosion*   [522-03 and 522-05]

The metal sheaths and armour of cables, metal conduit and conduit fittings, metal trunking and ducting, as well as the fixings of all these items, are likely to suffer corrosion in damp situations due to chemical or electrolytic attack by certain materials, unless special precautions are taken. The offending materials include:

1.   unpainted lime, cement and plaster,
2.   floors and dados including magnesium chloride,
3.   acidic woods, such as oak,
4.   plaster undercoats containing corrosive salts,
5.   dissimilar metals which will set up electrolytic action.

In all cases the solution to the problem of corrosion is to separate the materials between which the corrosion occurs. For chemical attack, this means having suitable coatings on the item to be installed, such as galvanising or an enamel or plastic coating. Bare copper sheathed cable, such as mineral insulated types, should not be laid in contact with galvanised material like a cable tray if conditions are likely to be damp. A pvc covering on the cable will prevent a possible corrosion probelm.

To prevent electrolytic corrosion, which is particularly common with aluminium-sheathed cables or conduit, a careful choice of the fixings with which the aluminium comes into contact is important, especially in damp situations. Suitable materials are aluminium, alloys of aluminium which are corrosion resistant, zinc alloys complying with BS 1004, porcelain, plastics, or galvanised or sheradised iron or steel

4.3   Cable choice

4.3.1   *Cable types* ~ [522-01 to 522-05, 523-03 and 527-01]

When choosing a cable one of the most important factors is the temperature attained by its insulation (see {4.1.1}); if the temperature is allowed to exceed the upper design value, premature failure is likely. In addition, corrosion of the sheaths or enclosures may result. For example, bare conductors such as busbars may be operated at much higher temperatures than most insulated conductors.

However, when an insulated conductor is connected to such a high temperature system, its own insulation may be affected by heat transmitted from the busbar, usually by conduction and by radiation. To ensure that the insulation is not damaged:
either the operating temperature of the busbar must not exceed the safe temperature for the insulation,
or the conductor insulation must be removed for a suitable distance from the connection with the busbar and replaced with heat resistant insulation (*see* {Fig 4.7}).

It is common sense that the cable chosen should be suitable for its purpose and for the surroundings in which it will operate. It should not be handled and installed in unsuitable temperatures. For example, p.v.c. becomes hard and brittle at low temp-eratures, and if a cable insulated with it is installed at temperatures below 5°C it may well become damaged.

[522] includes a series of Regulations which are intended to ensure that suitable cables are chosen to prevent damage from temperature levels, moisture, dust and dirt, pollution, vibration, mechanical stress, plant growths, animals, sunlight or the kind of building in which they are installed. As already mentioned in {3.5.2}, cables must not produce, spread, or sustain fire.

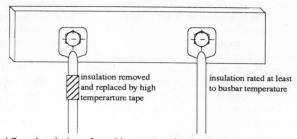

Fig 4.7   Insulation of a cable connected to hot busbar

BS 6387 covers cables which must be able to continue to operate in a fire. These special cables are intended to be used when it is required to maintain circuit integrity for longer than is possible with normal cables. Such cables are categorised with three letters. The first indicates the resistance to fire alone (A,B,C and S) and the second letter is a W and indicates that the cable will survive for a time at 650°C when also subject to water (which may be used to tackle the fire). The third letter (X, Y or Z) indicates the resistance to fire with mechanical shock. For full details of these special cables see the BS

4.3.2   *Current carrying capacity of conductors*   ~ [434-03-03, 523, 524 and Appendix 4]

All cables have electrical resistance, so there must be an energy loss when they carry current.This loss appears as heat and the temperature of the cable rises. As it does so, the heat it loses to its surroundings by conduction, convection and radiation also increases. The rate of heat loss is a function of the *difference* in temperature between the conductor and the surroundings, so as the conductor temperature rises, so does its rate of heat loss.

A cable carrying a steady current, which produces a fixed heating effect, will get hotter until it reaches the balance temperature where heat input is equal to heat loss {Fig 4.8}. The final temperature achieved by the cable will thus depend on the current carried, how easily heat is dissipated from the cable and the temperature of the cable surroundings.

P.V.C. is probably the most usual form of insulation, and is very susceptible to damage by high temperatures. It is very important that p.v.c. insulation should not be allowed normally to exceed 70°C, so the current ratings of cables are designed to ensure that this will not happen. Some special types of p.v.c. may be used up to 85°C, and higher temperatures (up to 160°C) are permitted under very short term fault conditions.

A different set of cable ratings will become necessary if the ability of a cable to shed its excess heat changes. Thus, [Appendix 4] has different Tables and columns for different types of cables, with differing conditions of installation, degrees of grouping and so on. For example, mineral insulation does not deteriorate, even at very high temperatures. The insulation is also an excellent heat conductor, so the rating of such a cable depends on how hot its sheath can become rather than the temperature of its insulation.

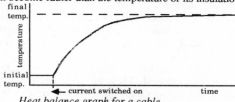

Fig 4.8   Heat balance graph for a cable

For example, if a mineral insulated cable has an overall sheath of LSF or p.v.c., the copper sheath temperature must not exceed 70°C, whilst if the copper sheath is bare and cannot be touched and is not in contact with materials which are combustible its temperature can be allowed to reach 105°C. Thus, a 1mm^2 light duty twin mineral insulated cable has a current rating of 18.5 A when it has an LSF or p.v.c. sheath, or 22 A if bare and not exposed to touch. It should be noticed that the cable volt drop will be higher if more current is carried (*see*{4.3.11}). [Appendix 4] includes a large number of Tables relating to the current rating of cables installed in various ways. The use of the Tables will be considered in more detail in {4.3.4 to 4.3.11}.

4.3.3   Methods of cable installation ~ [521-03, 521-07, 522-11, 522-12, 527-02 to 527-04, and 528-02]

We have seen that the rating of a cable depends on its ability to lose the heat produced in it by the current it carries and this depends to some extent on the way the cable is installed. A cable clipped to a surface will more easily be able to dissipate heat than a similar cable which is installed with others in a conduit.

[Table 4A] of [Appendix 4] lists twenty standard methods of installation, each of them taken into account in the rating tables of the same Appendix. For example, two 2.5 mm^2 single core p.v.c. insulated non-armoured cables drawn into a steel conduit (installation method 3) have a current rating of 24 A {Table 4.6}. A 2.5 mm^2 twin p.v.c. insulated and sheathed cable, which contains exactly the same conductors, has a current rating of 27 A {Table 4.7} when clipped directly to a non-metallic surface. Cables sheathed in p.v.c. must not be subjected to direct sunlight, because the ultra-violet component will leach out the plasticiser, causing the sheath to harden and crack. Cables must not be run in the same enclosure (*e.g.* trunking, pipe or ducting) as non-electrical services such as water, gas, air, *etc.* unless it has been established that the electrical system can suffer no harm as a result. If electrical and other services have metal sheaths and are touching, they must be bonded. Cables must not be run in positions where they may suffer or cause damage or interference with other systems. They should not, for example, be run alongside hot pipes or share a space with a hearing induction loop.

The build-up of dust on cables can act as thermal insulation. In some circumstances the dust may be flammable or even explosive. Design cable runs to minimise dust accumulation: run cables on vertically mounted cable ladders rather than horizontal cable trays. When cables are run together, each sets up a magnetic field with a strength depending on the current carried. This field surrounds other cables, so that there is the situation of current-carrying conductors situated in a magnetic field. This will result in a force on the conductor, which is usually negligible under normal conditions but which can become very high indeed when heavy currents flow under fault conditions. All cables and conductors must be properly fixed or supported to prevent damage to them under these conditions.

4.3.4   Ambient temperature correction factors ~ [523-01, and Appendix 4]

The transfer of heat, whether by conduction, convection or radiation, depends on temperature *difference* — heat flows from hot to cold at a rate which depends on the temperature difference between them. Thus, a cable installed near the roof of a boiler house where the surrounding (ambient) temperature is very high will not dissipate heat so readily as one clipped to the wall of a cold wine cellar.

[Appendix 4] includes two tables giving correction factors to take account of the ability of a cable to shed heat due to the ambient temperature. The Regulations use the symbol C_a to represent this correction factor. The tables assume that the ambient temperature is 30°C, and give a factor by which current rating is multiplied for other ambient temperatures.

For example, if a cable has a rating of 24 A and an ambient temperature correction factor of 0.77, the new current rating becomes 24 x 0.77 or 18.5 A. Different values are given depending on whether the circuit in question is protected by a semi-enclosed (rewirable) fuse or some other method of protection. The most useful of the correction factors are given in {Table 4.3}.

In {Table 4.3}, '70°C m.i.' gives data for mineral insulated cables with sheaths covered in p.v.c. or LSF or open to touch, and 105°C m.i.' for mineral insulated cables with bare sheaths which cannot be touched and are not in contact with combustible material. The cable which is p.v.c. sheathed or can be touched must run cooler than if it is bare and not in contact with combustible material, and so has lower correction factors.

Mineral insulated cables must have insulating sleeves in terminations with the same temperature rating as the seals used.

Where a cable is subjected to sunlight, it will not be able to lose heat so easily as one which is shaded. This is taken into account by adding 20°C to the ambient temperature for a cable which is unshaded.

Table 4.3 Correction factors for ambient temperature (C_a)
{from [Tables 4C1 and 4C2] of BS 7671: 1992)

Ambient temperature (°C)	Type of insulation			
	70°C p.v.c.	85°C rubber	70°C m.i.	105°C m.i.
25	1.03 (1.03)	1.02 (1.02)	1.03 (1.03)	1.02 (1.02)
30	1.00 (1.00)	1.00 (1.00)	1.00 (1.00)	1.00 (1.00)
35	0.94 (0.97)	0.95 (0.97)	0.93 (0.96)	0.96 (0.98)
40	0.87 (0.94)	0.90 (0.95)	0.85 (0.93)	0.92 (0.96)
45	0.79 (0.91)	0.85 (0.93)	0.77 (0.89)	0.88 (0.93)
50	0.71 (0.87)	0.80 (0.91)	0.67 (0.86)	0.84 (0.91)
55	0.61 (0.84)	0.74 (0.88)	0.57 (0.79)	0.80 (0.89)

Figures in brackets apply to semi-enclosed fuses used for overload protection

4.3.5　　Cable grouping correction factors　~　[Appendix 4]

If a number of cables is installed together and each is carrying current, they will all warm up. Those which are on the outside of the group will be able to transmit heat outwards, but will be restricted in losing heat inwards towards other warm cables. Cables 'buried' in others near the centre of the group may find it impossible to shed heat at all, and will rise further in temperature {Fig 4.9}.

Because of this, cables installed in groups with others (for example, if enclosed in a conduit or trunking) are allowed to carry less current than similar cables clipped to, or lying on, a solid surface which can dissipate heat more easily. If surface mounted cables are touching the reduction in the current rating is, as would be expected, greater than if they are separated. {Figure 4.9} illustrates the difficulty of dissipating heat in a group of cables.

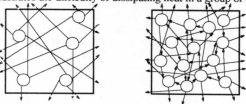

Fig 4.9　　The need for the grouping correction factor C_g
a) widely spaced cables dissipate heat easily
b) closely packed cables cannot dissipate heat and so their temperature rises

For example, if a certain cable has a basic current rating of 24 A and is installed in a trunking with six other circuits (*note carefully,* this is *circuits* and not *cables*), C_g has a value of 0.57 and the cable current rating becomes 24 x 0.57 or 13.7 A. The symbol C_g is used to represent the factor used for derating cables to allow for grouping. {Table 4.4} shows some of the more useful values of C_g.

The grouping factors are based on the assumption that all cables in a group are carrying rated current. If a cable is expected to carry no more than 30% of its grouped rated current, it can be ignored when calculating the group rating factor. For example, if there are four circuits in a group but one will be carrying less than 30% of its grouped rating, the group may be calculated on the basis of having only three circuits.

The grouping factor may also be applied to the determination of current ratings for cables as explained in {3.8}.

4.3.6   *Thermal insulation correction factors* ~   [523-04, Appendix 4]
The use of thermal insulation in buildings, in the forms of cavity wall filling, roof space blanketing, and so on, is now standard. Since the purpose of such materials is to limit the transfer of heat, they will clearly affect the ability of a cable to dissipate the heat build up within it when in contact with them.

The cable rating tables of [Appendix 4] allow for the reduced heat loss for a cable which is enclosed in an insulating wall and is assumed to be in contact with the insulation on one side. In all other cases, the cable should be fixed in a position where it is unlikely to be completely covered by the insulation. Where this is not possible and a cable is buried in thermal insulation for 0.5 m (500 mm) or more, a rating factor (the symbol for the thermal insulation factor is C_i) of 0.5 is applied, which means that the current rating is halved.

Table 4.4 Correction factors for groups of cables.
(from [Table 4B1] of BS 7671: 1992)

Number of circuits	Correction factor C_g		
	Enclosed or clipped	Clipped to non-metallic surface	
		Touching	Spaced*
2	0.80	0.85	0.94
3	0.70	0.79	0.90
4	0.65	0.75	0.90
5	0.60	0.73	0.90
6	0.57	0.72	0.90
7	0.54	0.72	0.90
8	0.52	0.71	0.90
9	0.50	0.70	0.90
10	0.48	———	0.90

* 'Spaced' means a gap between cables at least equal to cable diameter.

Table 4.5 Derating factors (C_i) for cables up to 10 mm² in cross-sectional area buried in thermal insulation.
(from [Table 52A] of BS 7671: 1992)

Length in insulation (mm)	Derating factor (C_i)
50	0.89
100	0.81
200	0.68
400	0.55
500 or more	0.50

If a cable is totally surrounded by thermal insulation for only a short length (for example, where a cable passes through an insulated wall), the heating effect on the insulation will not be so great because heat will be conducted from the short high-temperature length through the cable conductor. Clearly, the longer the length of cable enclosed in the insulation the greater will be the derating effect. {Table 4.5} shows the derating factors for lengths in insulation of up to 400 mm and applies to cables having cross-sectional area up to 10 mm².

Commonly-used cavity wall fillings, such as polystyrene sheets or granules, will have an adverse effect on p.v.c. sheathing, leeching out some of the plasticiser so that the p.v.c. becomes brittle. In such cases, an inert barrier must be provided to separate the cable from the thermal insulation. PVC cable in contact with bitumen may have some of its plasticiser removed: whilst this is unlikely to damage the cable, the bitumen will become fluid and may run.

4.3.7 *When a number of correction factors applies* ~ [523-01, Appendix 4]

In some cases all the correction factors will need to be applied because there are parts of the cable which are subject to all of them. For example, if a mineral insulated cable with p.v.c. sheath protected by a circuit breaker and with a tabulated rated current of 34 A is run within the insulated ceiling of a boiler house with an ambient temperature of 45°C and forms part of a group of four circuits, derating will be applied as follows:

Actual current rating (I_z)

= tabulated current (I_t) x ambient temperature factor(C_a) x group factor (C_g) x thermal insulation factor (C_i)

= 34 x 0.77 x 0.65 x 0.5 A = **8.5 A**

In this case, the current rating is only one quarter of its tabulated value due to the application of correction factors. A reduction of this sort will only occur when *all* the correction factors apply at the same time. There are many cases where this is not so. If, for example, the cable above were clipped to the ceiling of the boiler house and not buried in thermal insulation, the thermal insulation factor would not apply.

Then, I_z = I_t x C_a x C_g = 34 x 0.77 x 0.65 A = **17.0 A**

The method is to calculate the overall factor for each set of cable conditions and then to use the lowest only. For example, if on the way to the boiler house the cable is buried in thermal insulation in the wall of a space where the temperature is only 20°C and runs on its own, not grouped with other circuits, only the correction factor for thermal insulation would apply. However, since the cable is then grouped with others, and is subject to a high ambient temperature, the factors are:

C_i = 0.5
C_a x C_g = 0.77 x 0.65 = 0.5

The two factors are the same, so either (but not both) can be applied. Had they been different, the smaller would have been used.

4.3.8 *Protection by semi-enclosed (rewirable) fuses* ~ [433-02, Appendix 4]

If the circuit concerned is protected by a semi-enclosed (rewirable) fuse the cable size will need to be larger to allow for the fact that such fuses are not so certain in operation as are cartridge fuses or circuit breakers. The fuse rating must never be greater than 0.725 times the current carrying capacity of the lowest-rated conductor protected.

In effect, this is the same as applying a correction factor of 0.725 to all circuits protected by semi-enclosed (rewirable) fuses. The ambient temperature correction factors of {Table 4.3} are larger than those for other protective devices to take this into account.

4.3.9 Cable rating calculation ~ [433, 434, 523 and Appendix 4]
The Regulations indicate the following symbols for use when selecting cables:

I_z	is the current carrying capacity of the cable in the situation where it is installed
I_t	is the tabulated current for a single circuit at an ambient temperature of $30^{\circ}C$
I_b	is the design current, the actual current to be carried by the cable
I_n	is the rating of the protecting fuse or circuit breaker
I_2	is the operating current for the fuse or circuit breaker (the current at which the fuse blows or the circuit breaker opens)
C_a	is the correction factor for ambient temperature
C_g	is the correction factor for grouping
C_i	is the correction factor for thermal insulation.

The correction factor for protection by a semi-enclosed (rewirable) fuse is not given a symbol but has a fixed value of 0.725.

Under all circumstances, the cable current carrying capacity must be equal to or greater than the circuit design current and the rating of the fuse or circuit breaker must be at least as big as the circuit design current. These requirements are common sense, because otherwise the cable would be overloaded or the fuse would blow when the load is switched on.

To ensure correct protection from overload, it is important that the protective device operating current (I_2) is not bigger than 1.45 times the current carrying capacity of the cable (I_z). Additionally, the rating of the fuse or circuit breaker (I_n) must not be greater than the the cable current carrying capacity (I_z). It is important to appreciate that the operating current of a protective device is always larger than its rated value. In the case of a back-up fuse, which is not intended to provide overload protection, neither of these requirements applies.

To select a cable for a particular application, take the following steps: (note that to save time it may be better first to ensure that the expected cable for the required length of circuit will not result in the maximum permitted volt drop being exceeded {4.3.11}).

1. Calculate the expected (design) current in the circuit (I_b)
2. Choose the type and rating of protective device (fuse or circuit breaker) to be used (I_n)
3. Divide the protective device rated current by the ambient temperature correction factor (C_a) if ambient temperature differs from $30^{\circ}C$
4. Further divide by the grouping correction factor (C_g)
5. Divide again by the thermal insulation correction factor (C_i)
6. Divide by the semi-enclosed fuse factor of 0.725 where applicable
7. The result is the rated current of the cable required, which must be chosen from the appropriate tables {4.6 to 4.9}.

Observe that one should *divide* by the correction factors, whilst in the previous subsection we were *multiplying* them. The difference is that here we start with the design current of the circuit and adjust it to take account of factors which will derate the cable. Thus, the current carrying capacity of the cable will be greater than the design current. In {4.3.7} we were calculating by how much the current carrying capacity was reduced due to application of correction factors.

{Tables 4.6 to 4.9} give current ratings and volt drops for some of the more commonly used cables and sizes. They are extracted from the Regulations tables shown in square brackets *e.g.* [4D1A]

The examples below will illustrate the calculations, but do not take account of volt drop requirements (*see* {4.3.11}).

Example 4.1

An immersion heater rated at 240 V, 3 kW is to be installed using twin with protective conductor p.v.c. insulated and sheathed cable. The circuit will be fed from a 15 A miniature circuit breaker type 2, and will be run for much of its 14 m length in a roof space which is thermally insulated with glass fibre. The roof space temperature is expected to rise to 50°C in summer, and where it leaves the consumer unit and passes through a 50 mm insulation-filled cavity, the cable will be bunched with seven others. Calculate the cross-sectional area of the required cable.

First calculate the design current (I_b).
$$I_b = \frac{P}{U} = \frac{3000}{240} \text{ A} = 12.5 \text{ A}$$

The ambient temperature correction factor is found from {Table 4.3} to be 0.71. The group correction factor is found from {Table 4.4} as 0.52. (The circuit in question is bunched with seven others, making eight in all).

The thermal insulation correction factor is already taken into account in the current rating table [4D2A ref. method 4] and need not be further considered. This is because we can assume that the cable in the roof space is in contact with the glass fibre but not enclosed by it. What we must consider is the point where the bunched cables pass through the insulated cavity. From {Table 4.5} we have a factor of 0.89.

The correction factors must now be considered to see if more than one of them applies to the same part of the cable. The only place where this happens is in the insulated cavity behind the consumer unit. Factors of 0.52 (C_g) and 0.89 (C_i) apply. The combined value of these (0.463), which is lower than the ambient temperature correction factor of 0.71, and will thus be the figure to be applied. Hence the required current rating is calculated:-

$$I_z = \frac{I_n}{C_g \times C_i} = \frac{15}{0.52 \times 0.89} \text{ A} = 32.4 \text{ A}$$

From {Table 4.7}, 6 mm^2 p.v.c. twin with protective conductor has a current rating of 32 A. This is not quite large enough, so 10 mm^2 with a current rating of 43 A is indicated. Not only would this add considerably to the costs, but would also result in difficulties due to terminating such a large cable in the accessories.

A more sensible option would be to look for a method of reducing the required cable size. For example, if the eight cables left the consumer unit in two bunches of four, this would result in a grouping factor of 0.65 (from {Table 4.4}). Before applying this, we must check that the combined grouping and thermal insulation factors (0.65 x 0.89 = 0.58) are still less than the ambient temperature factor of 0.71, which is the case.

This leads to a cable current rating of $\dfrac{15}{0.65 \times 0.89}$ A $= 25.9$ A .

This is well below the rating for **6 mm^2** of 32 A, so a cable of this size could be selected.

Table 4.6 Current ratings and volt drops for unsheathed single core p.v.c. insulated cables
(from [Tables 4D1A and 4D1B] of BS 7671: 1992)

Cross-sec area (mm²)	In conduit in thermal insulation (A) 2 cables	In conduit in thermal insulation (A) 3 or 4 cables	In conduit on wall (A) 2 cables	In conduit on wall (A) 3 or 4 cables	Clipped direct (A) 2 cables	Clipped direct (A) 3 or 4 cables	Volt drop (mV/A/m) 2 cables	Volt drop (mV/A) 3 or 4 cables
1.0	11.0	10.5	13.5	12.0	15.5	14.0	44.0	38.0
1.5	14.5	13.5	17.5	15.5	20.0	18.0	29.0	25.0
2.5	19.5	18.0	24.0	21.0	27.0	25.0	18.0	15.0
4.0	26.0	24.0	32.0	28.0	37.0	33.0	11.0	9.5
6.0	34.0	31.0	41.0	36.0	47.0	43.0	7.3	6.4
10.0	46.0	42.0	57.0	50.0	65.0	59.0	4.4	3.8
16.0	61.0	56.0	76.0	68.0	87.0	79.0	2.8	2.4

Table 4.7 Current ratings and volt drops for sheathed multi-core p.v.c.-insulated cables
(from [Tables 4D2A and 4D2B] of BS 7671: 1992)

Note that the classification "3 or 4 core" refers to three-phase circuits and not to applications such as two way switching.

Cross sectional area (mm²)	In wall in thermal insulation (A) 2 core	In wall in thermal insulation (A) 3 or 4 core	In conduit on wall (A) 2 core	In conduit on wall (A) 3 or 4 core	Clipped direct (A) 2 core	Clipped direct (A) 3 or 4 core	Volt drop (mV/A/m) 2 core	Volt drop (mV/A/m) 3 or 4 core
1.0	11.0	10.0	13.0	11.5	15.0	13.5	44.0	38.0
1.5	14.0	13.0	16.5	15.0	19.5	17.5	29.0	25.0
2.5	18.5	17.5	23.0	20.0	27.0	24.0	18.0	15.0
4.0	25.0	23.0	30.0	27.0	36.0	32.0	11.0	9.5
6.0	32.0	29.0	38.0	34.0	46.0	41.0	7.3	6.4
10.0	43.0	39.0	52.0	46.0	63.0	57.0	4.4	3.8
16.0	57.0	52.0	69.0	62.0	85.0	76.0	2.8	2.4

Table 4.8 Current ratings of mineral insulated cables clipped direct
(from [Tables 4I1A and 4I2A] of the 16th edition of the IEE Wiring Regulations)

Cross-sectional area (mm²)	p.v.c. sheath 2 x single or twin (A)	p.v.c. sheath 3 core (A)	p.v.c. sheath 3 x single or twin (A)	bare sheath 2 x single (A)	bare sheath 3 x single (A)
1.0 (500V)	18.5	16.5	16.5	22.0	21.0
1.5 (500V)	24.0	21.0	21.0	28.0	27.0
2.5 (500V)	31.0	28.0	28.0	38.0	36.0
4.0 (500V)	42.0	37.0	37.0	51.0	47.0
1.0 (750V)	20.0	17.5	17.5	24.0	24.0
1.5 (750V)	25.0	22.0	22.0	31.0	30.0
2.5 (750V)	34.0	30.0	30.0	42.0	41.0
4.0 (750V)	45.0	40.0	40.0	55.0	53.0
6.0 (750V)	57.0	51.0	51.0	70.0	67.0
10.0 (750V)	78.0	69.0	69.0	96.0	91.0
16.0 (750V)	104.0	92.0	92.0	127.0	119

Note that in {Tables 4.8 and 4.9} 'P.V.C. sheath' means bare and exposed to touch or having an overall covering of p.v.c.or LSF and 'Bare' means bare and neither exposed to touch nor in contact with combustible materials.

Table 4.9 Volt drops for mineral insulated cables
(from [Tables 4J1B and 4J2B] of BS 7671: 1992)

Cross-sectional area (mm²)	Single-phase p.v.c. sheath (mV/A/m)	Single-phase bare (mV/A/m)	Three-phase p.v.c. sheath (mV/A/m)	Three-phase bare (mV/A/m)
1.0	42.0	47.0	36.0	40.0
1.5	28.0	31.0	24.0	27.0
2.5	17.0	19.0	14.0	16.0
4.0	10.0	12.0	9.1	10.0
6.0	7.0	7.8	6.0	6.8
10.0	4.2	4.7	3.6	4.1
16.0	2.6	3.0	2.3	2.6

Example 4.2

The same installation as in Example 4.1 is proposed. To attempt to make the cable size smaller, the run in the roof space is to be kept clear of the glass fibre insulation. Does this make any difference to the selected cable size?

There is no correction factor for the presence of the glass fibre, so the calculation of I_Z will be exactly the same as Example 4.1 at 32.4 A.

This time reference method 1 (clipped direct) will apply to the current rating {Table 4.7}. For a two core cable, **4.0 mm²**, two core has a rating of 36 A, so this will be the selected size.

It is of interest to notice how quite a minor change in the method of installation, in this case clipping the cable to the joists or battens clear of the glass fibre, has reduced the acceptable cable size.

Example 4.3

Assume that the immersion heater indicated in the two previous examples is to be installed, but this time with the protection of a 15 A rewirable (semi-enclosed) fuse. Calculate the correct cable size for each of the alternatives, that is where firstly the cable is in contact with glass fibre insulation, and secondly where it is held clear of it.

This time the value of the acceptable current carrying capacity I_Z will be different because of the need to include a factor for the rewirable fuse as well as the new ambient temperature and grouping factors for the rewirable fuse from {Tables 4.3 and 4.4}.

$$I_Z = \frac{I_n}{C_g \times C_a \times 0.725} = \frac{15}{0.52 \times 0.87 \times 0.725} \text{ A} = 45.7 \text{ A}$$

In this case, the cable is in contact with the glass fibre, so the first column of {Table 4.7} of current ratings will apply. The acceptable cable size is 16 mm² which has a current rating of 57 A.

This cable size is not acceptable on the grounds of high cost and because the conductors are likely to be too large to enter the connection tunnels of the immersion heater and its associated switch. If the cables leaving the consumer unit are re-arranged in two groups of four, this will reduce the grouping factor to 0.65, so that the newly calculated value of I_Z is 36.6 A. This means using 10 mm² cable with a current rating of 43 A (from {Table 4.7}), since 6 mm² cable is shown to have a current rating in these circumstances of only 32 A. By further rearranging the cables leaving the consumer unit to be part of a group of only two, Cg is increased to 0.8, which reduces Iz to 29.7 A which enables selection of a **6 mm²** cable.

Should it be possible to bring the immersion heater cable out of the consumer unit on its own, no grouping factor would apply and Iz would fall to 23.8 A, allowing a 4 mm² cable to be selected.

Where the cable is not in contact with glass fibre there will be no need to repeat the calculation of I_Z, which still has a value of 29.7 A provided that it is possible to group the immersion heater cable with only one other where it leaves the consumer unit.. This time we use the 'Reference Method 1 (clipped direct)' column of the current rating {Table 4.7}, which shows that **4 mm²** cable with a current rating of 36 A will be satisfactory.

Examples 4.1, 4.2 and 4.3 show clearly how forward planning will enable a more economical and practical cable size to be used than would appear necessary at first. It is, of course, important that the design calculations are recorded and retained in the installation manual.

Example 4.4

A 415 V 50 Hz three-phase motor with an output of 7.5 kW, power factor 0.8 and efficiency 85% is the be wired using 500 V light duty three-core mineral insulated p.v.c. sheathed cable. The length of run from the HBC protecting fuses is 20 m, and for about half this run the cable is clipped to wall surfaces. For the remainder it shares a cable tray, touching two similar cables across the top of a boiler room where the ambient temperature is 50°C. Calculate the rating and size of the correct cable.

The first step is to calculate the line current of the motor.

$$\text{Input} = \frac{\text{output}}{\text{efficiency}} = \frac{7.5 \times 100}{85} \text{ kW} = 8.82 \text{ kW}$$

$$\text{Line current } I_b = \frac{P}{\sqrt{3} \times U_L \times \cos\phi} = \frac{8.82 \times 10^3}{\sqrt{3} \times 415 \times 0.8} \text{ A} = \textbf{15.3 A}$$

We must now select a suitable fuse. {Fig 3.15} for BS 88 fuses shows the 16 A size to be the most suitable. Part of the run is subject to an ambient temperature of 50°C, where the cable is also part of a group of three, so the appropriate correction factors must be applied. from {Tables 4.3 and 4.4}.

$$I_z = \frac{I_n}{C_g \times C_a} = \frac{16}{0.70 \times 0.67} \text{ A} = 34.2A$$

Note that the grouping factor of 0.70 has been selected because where the cable is grouped it is clipped to a metallic cable tray, and not to a non-metallic surface. Next the cable must be chosen from {Table 4.8}. Whilst the current rating would be 15.3 A if all of the cable run were clipped to the wall, part of the run is subject to the two correction factors, so a rating of 34.2 A must be used. For the clipped section of the cable (15.3 A), reference method 1 could be used which gives a size of 1.0 mm² (current rating 16.5 A). However, since part of the cable is on the tray (method 3) the correct size for 34.2 A will be **4.0 mm²**, with a rating of 37 A.

4.3.10 Special formulas for grouping factor calculation ~ [App. 4]

In some cases the conductor sizes where cables are grouped and determined by the methods shown in {4.3.9} can be reduced by applying some rather complex formulas given in [Appendix 4].

Whilst it is true that in many cases the use of these formulas will show the installer that it is safe for him to use a smaller cable than he would have needed by simple application of correction factors, this is by no means always the case. There are many cases where their application will make no difference at all. Since this book is for the electrician, rather than for the designer, the rather complicated mathematics will be omitted.

4.3.11 Cable volt drop ~ [525, Appendix 4]

All cables have resistance, and when current flows in them this results in a volt drop. Hence, the voltage at the load is lower than the supply voltage by the amount of this volt drop.

The volt drop may be calculated using the basic Ohm's law formula

$$U = I \times R$$

where U is the cable volt drop (V)

I is the circuit current (A), and

R is the circuit resistance (Ω)

Unfortunately, this simple formula is seldom of use in this case, because the cable resistance under load conditions is not easy to calculate.

[525-01-03] indicates that the voltage at any load must never fall so low as to impair the safe working of that load, or fall below the level indicated by the relevant British Standard where one applies.

[525-01-02] indicates that these requirements will be met if the voltage drop does not exceed 4% of the declared supply voltage. If the supply is single-phase at the usual level of 240 V, this means a maximum volt drop of 4% of 240 V which is 9.6 V, giving (in simple terms) a load voltage as low as 230.4 V. For a 415 V three-phase system, allowable volt drop will be 16.6 V with a line load voltage as low as 398.4 V.

It should be borne in mind that European Agreement HD 472 S2 allows the declared supply voltage of 230 V to vary by +10% or -6%. Assuming that the supply voltage of 240 V is 6% low, and allowing a 4% volt drop, this gives permissible load voltages of 216.6 V for a single-phase supply, or 374.5 V (line) for a 415 V three-phase supply.

To calculate the volt drop for a particular cable we use {Tables 4.6, 4.7 and 4.9}. Each current rating table has an associated volt drop column or table. For example, multicore sheathed non-armoured p.v.c. insulated cables are covered by {Table 4.7} for current ratings, and volt drops. The exception in the Regulations to this layout is for mineral insulated cables where there are separate volt drop tables for single- and three-phase operation, which are combined here as {Table 4.9}.

Each cable rating in the Tables of [Appendix 4] has a corresponding volt drop figure in millivolts per ampere per metre of run (mV/A/m). Strictly this should be mV/(A m), but here we shall follow the pattern adopted by the Wiring Regulations. To calculate the cable volt drop:

1. take the value from the volt drop table (mV/A/m)
2. multiply by the actual current in the cable (NOT the current rating)
3. multiply by the length of run in metres
4. divide the result by one thousand (to convert millivolts to volts).

For example, if a 4 mm^2 p.v.c. sheathed circuit feeds a 6 kW shower and has a length of run of 16 m, we can find the volt drop thus:

From {Table 4.7}, the volt drop figure for 4 mm^2 two-core cable is 11 mV/A/m.

Cable current is calculated from $I = \dfrac{P}{U} = \dfrac{6000}{240}$ A = 25 A

Volt drop is then $\dfrac{11 \times 25 \times 16}{1000}$ V = 4.4 V

Since the permissible volt drop is 4% of 240 V, which is 9.6 V, the cable in question meets volt drop requirements. The following examples will make the method clear.

Example 4.5

Calculate the volt drop for the case of Example 4.1. What maximum length of cable would allow the installation to comply with the volt drop regulations?

The table concerned here is {4.7}, which shows a figure of 7.3 mV/A/m for 6 mm^2 twin with protective conductor pvc insulated and sheathed cable. The actual circuit current is 12.5 A, and the length of run is 14 m.

Volt drop $= \dfrac{7.3 \times 12.5 \times 14}{1000}$ V = **1.28 V**

Maximum permissible volt drop is 4% of 240 V $= \dfrac{4 \times 240}{100}$ V = 9.6 V

If a 14 m run gives a volt drop of 1.28 V, the length of run for a 9.6 V drop will be $\dfrac{9.6}{1.28} \times 14$ m = **105 m**

Example 4.6
Calculate the volt drop for the case of {Example 4.2}. What maximum length of cable would allow the installation to comply with the volt drop regulations?

The Table concerned here is {4.7} which shows a volt drop figure for 4.0 mm^2 cable of 11 mV/A/m, with the current and the length of run remaining at 12.5 A and 14 m respectively.

Volt drop = $\dfrac{11 \times 12.5 \times 14}{1000}$ V = **1.93 V**

Maximum possible volt drop is 4% of 240 V = $\dfrac{4}{100}$ x 240 V = 9.6 V

If a 14 m run gives a volt drop of 1.93 V, the length of run for a 9.6 V drop will be
$\dfrac{9.6 \times 14}{1.93}$ m = **70 m**

Example 4.7
Calculate the volt drop for the cases of {Example 4.3} for each of the alternative installations. What maximum length of cable would allow the installation to comply with the volt drop regulations in each case?

In neither case is there any change in cable sizes, the selected cables being 6 mm^2 in the first case and 4 mm^2 in the second. Solutions are thus the same as those in {Examples 4.5 and 4.6} respectively.

Example 4.8
Calculate the volt drop and maximum length of run for the motor circuit of {Example 4.4}.

This time we have a mineral insulated p.v.c. sheathed cable, so volt drop figures will come from {Table 4.9}. This shows 9.1 mV/A/m for the 4 mm^2 cable selected, which must be used with the circuit current of 15.3 A and the length of run which is 20 m.

Volt drop will be $\dfrac{9.1 \times 15.3 \times 20}{1000}$ V = **2.78 V**

Permissible maximum volt drop is 4% of 415 V or $\dfrac{4 \times 415}{100}$ V = 16.6 V

Maximum length of run for this circuit with the same cable size and type will be
$\dfrac{16.6 \times 20}{2.78}$ m = **119 m**

The 'length of run' calculations carried out in these examples are often useful to the electrician when installing equipment at greater distances from the mains position.

It is important to appreciate that the allowable volt drop of 4% of the supply voltage *applies to the whole of an installation*. If an installation has mains, submains and final circuits, for instance, the volt drop in each must be calculated and added to give the total volt drop as indicated in {Fig 4.10}. All of our work in this sub-section so far has assumed that cable resistance is the only factor responsible for volt drop. In fact, larger cables have significant self inductance as well as resistance. As we shall see in Chapter 5 there is also an effect called *impedance* which is made up of resistance and inductive reactance (*see* {Fig 5.8(a)}).

Inductive reactance $\qquad X_L = 2\pi f L$
where $\qquad X_L =$ inductive reactance in ohms (Ω)
$\qquad\qquad \pi =$ the mathematical constant 3.142
$\qquad\qquad f =$ the system frequency in hertz (Hz)
$\qquad\qquad L =$ circuit self inductance in henrys (H)

It is clear that inductive reactance increases with frequency, and for this reason the volt drop tables apply only to systems with a frequency lying between 49 Hz and 61 Hz.

For small cables, the self inductance is such that the inductive reactance, is small compared with the resistance. Only with cables of cross-sectional area 25 mm^2 and greater need reactance be considered. Since cables as large as this are seldom used on work which has not been designed by a qualified engineer, the subject of reactive volt drop component will not be further considered here.

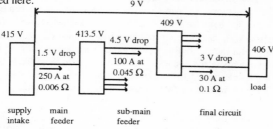

Fig 4.10   Total volt drop in large installations

If the actual current carried by the cable (the design current) is less than the rated value, the cable will not become as warm as the calculations used to produce the volt drop tables have assumed. The Regulations include (in [Appendix 4]) a very complicated formula to be applied to cables of cross-sectional area 16 mm^2 and less which may show that the actual volt drop is less than that obtained from the tables. This possibility is again seldom of interest to the electrician, and is not considered here.

4.3.12   *Harmonic currents and neutral conductors*   ~ [331-01, 473-03-03 to 473-03-04, and 524-02]

A perfectly balanced three-phase system (one with all three phase loads identical in all respects) has no neutral current and thus has no need of a neutral conductor. This is often so with motors, which are fed through three core cables in most cases.

Many three-phase loads are made up of single-phase loads, each connected between one line and neutral. It is not likely in such cases that the loads will be identical, so the neutral will carry the out-of-balance current of the system. The greater the degree of imbalance, the larger the neutral current.

Some three-phase four-core cables have a neutral of reduced cross-section on the assumption that there will be some degree of balance. Such a cable must not be used unless the installer is certain that severe out-of-balance conditions will never occur. Similar action must be taken with a three-phase circuit wired in single-core cables. A reduced neutral conductor may only be used where out-of-balance currents will be very small compared to the line currents.

A problem is likely to occur in systems which generate significant third harmonic currents. Devices such as discharge lamp ballasts and transformers on low load distort the current waveform. Thus, currents at three times normal frequency (third harmonics) are produced, which do not cancel at the star point of a three-phase system as do normal frequency currents, but add up, so that the neutral carries very heavy third harmonic currents. For this reason, it is important not to reduce the cross-sectional area of a neutral used to feed discharge lamps (including fluorescent lamps).

In some cases it may be necessary to insert overload protection in a neutral conductor. Such protection must be arranged to open all phase conductors on operation, but *not the neutral*. This clearly indicates the use of a special circuit breaker.

It is very important that the neutral of each circuit is kept quite separate from those of other circuits. Good practice suggests that the separate circuit neutrals should be connected in the same order at the neutral block as the corresponding phase conductors at the fuses or circuit breakers.

4.3.13 *Low smoke-emitting cables* [527-01]

Normal p.v.c. insulation emits dense smoke and corrosive gases when burning. If cables are to be run in areas of public access, such as schools, supermarkets, hospitals, *etc*, the designer should consider the use of special cables such as those with thermo-setting or elastomeric insulation which do not cause such problems in the event of fire. This action is most likely to be necessary in areas expected to be crowded, along fire escape routes, and where equipment is likely to suffer damage due to corrosive fumes.

4.3.14 *The effects of animals, insects and plants*

Cables may be subject to damage by animals and plants as well as from their environment. Rodents in particular seem to have a particular taste for some types of cable sheathing and can gnaw through sheath and insulation to expose the conductors. Cables impregnated with repellant chemicals are not often effective and may also fall foul of the Health and Safety Regulations. Rodents build nests, often of flammable materials, leading to a fire hazard. Care should be taken to avid cable installation along possible vermin runs, but where this cannot be avoided, steel conduit may be the answer.

Mechanical damage to wiring systems by larger animals such as cattle and horses can often be prevented by careful siting of cable runs and outlets. Attention must also be given to the fact that waste products from animals may be corrosive. Access by insects is difficult to prevent. but vent holes can be sealed with breathers. Damage by plants is a possible hazard, the effect of tree roots on small lighting columns being an obvious problem area.

4.4 Cable supports, joints and terminations.

4.4.1 *Cable supports and protection* ~ [521-03, 522-06-04 to 522-06-07, 522-08-04 to 522-08-06]

Cables must be fixed securely at intervals which are close enough to ensure that there will be no excessive strain on the cable or on its joints and terminations, and to prevent cable loops appearing which could lead to mechanical damage. {Table 4.10} indicates minimum acceptable spacings of fixings for some common types of cables.

Table 4.10 Maximum spacing for cable supports

Overall cable diameter (mm)	p.v.c. sheathed		Mineral insulated	
	horizontal (mm)	vertical (mm)	horizontal (mm)	vertical (mm)
up to 9	250	400	600	800
10 to 15	300	400	900	1200
16 to 20	350	450	1500	2000
21 to 40	400	550	2000	3000

Where cable runs are neither vertical nor horizontal, the spacing depends on the angle as shown in {Fig 4.11}.

Where a cable is flat in cross-section as in the case of a p.v.c. insulated and sheathed type, the overall diameter is taken as the major axis as shown in {Fig 4.12}.

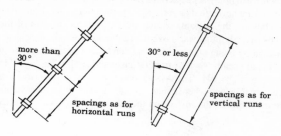

Fig 4.11   Spacing of support clips on angled runs

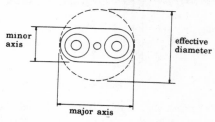

Fig 4.12   Effective diameter of a flat cable

The Regulations are concerned to protect hidden cables from damage. Thus, where cables are run beneath boarded floors, they must pass through holes drilled in the joists which are at least 50 mm below the top surface of the joist. This is to prevent accidental damage due to nails being driven into the joists. The hole diameters must not exceed one quarter of the depth of the joist and they must be drilled at the joist centre (the neutral axis). Hole centres must be at least three diameters apart, and the holes must only be drilled in a zone which extends 25% to 40% of the beam length from both ends.

An alternative is to protect the cable in steel conduit. It is not practicable to thread rigid conduit through holes in the joists, so the steel conduit may be laid in slots cut in the upper or lower edges as shown in {Fig 4.13}. The depth of the slot must be no greater than one eighth of the joist depth and notches must be in a zone extending from 10% to 25% of the beam length from both ends.

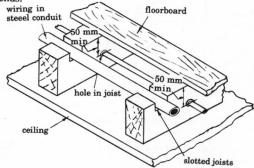

Fig 4.13   Support and protection for cables run under floors

Where cable runs are concealed behind plaster they must be installed in 'acceptable zones' which are intended to reduce the danger to the cables and to people who drill holes or knock nails into walls. Cable runs must only follow paths which are horizontal or vertical from an outlet, or be within 150 mm of the top (but not the bottom) of the wall, or within 150 mm of the angle formed by two adjoining walls. Where a cable run has to be diagonal, it must be protected by being enclosed in steel conduit, or must be a cable with an earthed metal sheath (such as mineral insulated cable). The possible zones are shown in {Fig 4.14}. The internal partition walls of some modern buildings are very thin, and where cables complying with the requirements above are within 50 mm of the surface on the other side, they will require protection.

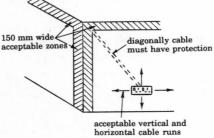

150 mm wide acceptable zones

diagonally cable must have protection

acceptable vertical and horizontal cable runs

Fig 4.14   Acceptable installation zones for concealed cables.
The diagonal cable must be enclosed in earthed metal

There are cases where cables are enclosed in long vertical runs of trunking or conduit. The weight of the cable run, which effectively is hanging onto the top support, can easily cause damage by compressing the insulation where it is pulled against the support. In trunking there must be effective supports no more than 5 m apart, examples of which are shown in Fig 4.15, whilst for conduit the run must be provided with adaptable boxes at similar intervals which can accommodate the necessary supports.

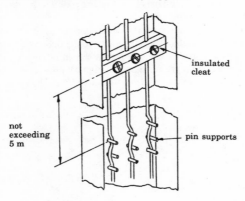

insulated cleat

not exceeding 5 m

pin supports

Fig 4.15   Support for vertical cables in trunking

The top of a vertical conduit or trunking run must have a rounded support to reduce compression of insulation. The diameters required will be the same as those for cable bends given in {4.4.2}.

Support for overhead conductors is considered in {7.13}.

4.4.2 Cable bends ~ [522-08]

If an insulated cable is bent too sharply, the insulation and sheath on the inside of the bend will be compressed, whilst that on the outside will be stretched. This can result in damage to the cable as shown in {Fig 4.16}.

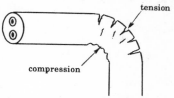

Fig 4.16 Damage to cable insulation due to bending

The bending factor must be used to assess the minimum acceptable bending radius, values for common cables being given in {Table 4.11}.

Table 4.11 Bending Factors for common cables		
Type of insulation	*Overall diameter*	*Bending factor*
p.v.c.	up to 10 mm	3 (2)
p.v.c.	10 mm to 25 mm	4 (3)
p.v.c.	over 25 mm	6
mineral	any	6 *
The figures in brackets apply to unsheathed single-core stranded p.v.c. cables when installed in conduit, trunking or ducting.		

*Mineral insulated cables may be bent at a minimum radius of three times cable diameter provided that they will only be bent once. This is because the copper sheath will 'work harden' when bent and is likely to crack if straightened and bent again.

The factor shown in the table is that by which the overall cable diameter {Fig 4.12} must be multiplied to give the minimum inside radius of the bend. For example, 2.5 mm² twin with protective conductor sheathed cable has a cross-section 9.7 mm x 5.4 mm. Since the Table shows a factor of 3 for this size, the minimum inside radius of any bend must be 3 x 9.7 = 29.1 mm.

4.4.3   Joints and terminations ~ [526]

The normal installation has many joints, and it follows that these must all remain safe and effective throughout the life of the system. With this in mind, regulations on joints include the following:

1. All joints must be durable, adequate for their purpose, and mechanically strong.

2. They must be constructed to take account of the conductor material and insulation, as well as temperature: *eg*, a soldered joint must not be used where the temperature may cause the solder to melt or to weaken. Very large expansion forces are not uncommon in terminal boxes situated at the end of straight runs of large cables when subjected to overload or to fault currents.

3. All joints and connections must be made in an enclosure complying with the appropriate British Standard.

4. Where sheathed cables are used, the sheath must be continuous into the joint enclosure {Figure 4.17}.

5. All joints must be accessible for inspection and testing unless they are buried in compound or encapsulated, are between the cold tail and element of a heater such as a pipe tracer or underfloor heating system, or are made by soldering, welding, brazing or compression.

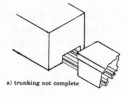 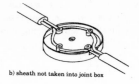

a) trunking not complete   b) sheath not taken into joint box

Fig 4.17   Failure to enclose non-sheathed cables

4.5   Cable enclosures

4.5.1   *Plastic and metal conduits*   ~   *[521-04, 522-03-01 and 522-03-02, 522-05, 522-06-01 and 522-06-02, 522-09, 522-10, 527-02 and 527-03]*

A system of conduits into which unsheathed cables can be drawn has long been a standard method for electrical installations. The Regulations applying to conduit systems may be summarised as follows:

1. All conduits and fittings must comply with the relevant British Standards.

2. Plastic conduits must not be used where the ambient temperature or the temperature of the enclosed cables will exceed 60°C. Cables with thermo-setting insulation are permitted to run very hot, and must be suitably down-rated when installed in plastic conduit. To prevent the spread of fire, plastic conduits (and plastic trunking) must comply with ignitability characteristic 'P' of BS 476 Part 5.

3. Conduit systems must be designed and erected so as to exclude moisture, dust and dirt. This means that they must be completely closed, with box lids fitted. To ensure that condensed moisture does not accumulate, small drainage holes must be provided at the lowest parts of the system.

4. Proper precautions must be taken against the effects of corrosion (*see* {4.2.5}), as well as against the effects of flora (plant growths) and fauna (animals). Protection from rusting of steel conduit involves the use of galvanised (zinc coated) tubing, and against electrolytic corrosion the prevention of contact between dissimilar metals *eg* steel and aluminium. Any additional protective conductor must be run *inside* the conduit or its reactance is likely to be so high that it becomes useless if intended to reduce fault loop impedance.

5. A conduit system must be completely erected before cables are drawn in. It must be free of burrs or other defects which could damage cables whilst being inserted.

6. The bends in the system must be such that the cables drawn in will comply with the minimum bending radius requirements {4.4.2}.

7. The conduit must be installed so that fire cannot spread through it, or through holes cut in floors or walls to allow it to pass. This subject of fire spread will be considered in greater detail in {4.5.2}.

8. Allowance must be made, in the form of expansion loops, for the thermal expansion of long runs of metal or plastic conduit. Remember that plastic expands and contracts more than steel.

9   Use flexible joints when crossing building expansion joints

Table 4.12   Maximum spacing of supports for conduits

Conduit diameter (mm)	Rigid metal (m) Horizontal	Vertical	Rigid insulating (m) Horizontal	Vertical
up to 16	0.75	1.0	0.75	1.0
16 to 25	1.75	2.0	1.5	1.75
25 to 40	2.0	2.25	1.75	2.0
over 40	2.25	2.5	2.0	2.0

4.5.2 **Ducting and trunking** ~ [521-01-02, 521-05, 521-06, 521-07-01 and 521-07-02, 522-01-01, 522-03-01 and 522-03-02, 522-04 and 522-05, 527-03 and 527-04]

Metal and plastic trunkings are very widely used in electrical installations. They must be manufactured to comply with the relevant British Standards, and must be installed so as to ensure that they will not be damaged by water or by corrosion (*see* {4.2.5}).

Table 4.13 Support spacings for trunking				
Maximum distances between supports are in metres.				
Typical trunking size	Metal		Insulating	
(mm)	Horizontal	Vertical	Horizontal	Vertical
up to 25 x 25	0.75	1.0	0.5	0.5
up to 50 x 25	1.25	1.5	0.5	0.5
up to 50 x 50	1.75	2.0	1.25	1.25
up to 100 x 50	3.0	3.0	1.75	2.0

If it is considered necessary to provide an additional protective conductor in parallel with steel trunking, it must be run *inside* the trunking or the presence of steel between the live and protective cables will often result in the reactance of the protective cable being so high that it will have little effect on fault loop impedance. Trunking must be supported as indicated in {Table 4.13}. The table does not apply to special lighting trunking which is provided with strengthened couplers. Where crossing a building expansion joint a suitable flexible joint should be included.

Where trunking or conduit passes through walls or floors the hole cut must be made good after the first fix on the construction site to give the partition the same degree of fire protection it had before the hole was cut. Since it is possible for fire to spread through the interior of the trunking or conduit, fire barriers must be inserted as shown in {Fig 4.18}. An exception is conduit or trunking with a cross-sectional area of less than 710 mm^2, so that conduits up to 32 mm in diameter and trunking up to 25 mm x 25 mm need not be provided with fire barriers. During installation, temporary fire barriers must be provided so that the integrity of the fire prevention system is always maintained.

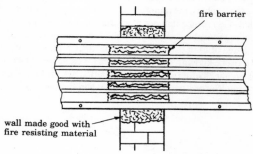

fire barrier

wall made good with fire resisting material

Fig 4.18 *Provision of fire barriers in ducts and trunking*

Since trunking will not be solidly packed with cables (*see* {4.5.3}) there will be room for air movement. A very long vertical trunking run may thus become extremely hot at the top as air heated by the cables rises; this must be prevented by barriers as shown in {Fig 4.19}. In many cases the trunking will pass through floors as it rises, and the fire stop barriers needed will also act as barriers to rising hot air.

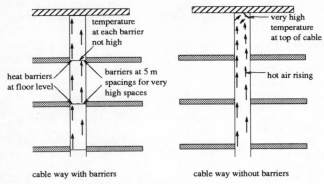

cable way with barriers   cable way without barriers

Fig 4.19   Heat barriers provided in vertical cable ways

Lighting trunking is being used to a greater extent than previously. In many cases, it includes copper conducting bars so that luminaires can be plugged in at any point, especially useful for display lighting.

The considerably improved life, efficiency and colour rendering properties of extra-low voltage tungsten halogen lamps has led to their increasing use, often fed by lighting trunking. It is important here to remember that whilst the voltage of a 12 V lamp is only one twentieth of normal mains potential, the current for the same power inputs will be twenty times greater. Thus, a trunking feeding six 50 W 12 V lamps will need to be rated at 25 A.

4.5.3   Cable capacity of conduits and trunking

Not only must it be possible to draw cables into completed conduit and trunking systems, but neither the cables nor their enclosures must be damaged in the process. If too many cables are packed into the space available, there will be a greater increase in temperature during operation than if they were given more space. It is important to appreciate that grouping factors (*see* {4.3.5}) still apply to cables enclosed in conduit or trunking.

To calculate the number of cables which may be drawn into a conduit or trunking, we make use of four tables ({Tables 4.14 to 4.17}). For situations not covered by these tables, the requirement is that a space factor of 45% must not be exceeded. This means that not more than 45% of the space within the conduit or trunking must be occupied by cables, and involves calculating the cross-sectional area of each cable, including its insulation, for which the outside diameter must be known. The cable factors for cables with thermosetting insulation are higher than those for pvc insulation when the cables are installed in trunking, but the two are the same when drawn into conduit *(see* {Table 4.14})

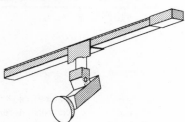

Fig 4.20   Low voltage luminaire on lighting trunking

The figures in {Table 4.14} may be high when applied to some types of plastic trunking due to the large size of the internal lid fixing clips.

Table 4.14 Cable factors (terms) for conduit and trunking

Type of conductor	Conductor c.s.a. (mm²)	Factor for conduit	Factor for trunking pvc insulation	Factor for trunking thermosetting insulation
solid	1.0	16	3.6	3.8
solid	1.5	22	8.0	8.6
stranded	1.5	22	8.6	9.1
solid	2.5	30	11.9	11.9
stranded	2.5	30	12.6	13.9
stranded	4.0	43	16.6	18.1
stranded	6.0	58	21.2	22.9
stranded	10.0	105	35.3	36.3
stranded	16.0	145	47.8	50.3
stranded	25.0	217	73.9	75.4

Table 4.15 Cable factors (terms) for short straight runs up to 3 m

Type of conductor	Conductor c.s.a. (mm²)	Cable factor
solid	1.0	22
solid	1.5	27
solid	2.5	39
stranded	1.5	31
stranded	2.5	43
stranded	4.0	58
stranded	6.0	88
stranded	10.0	146

Table 4.16 Conduit factors (terms)

Length of run between boxes (m)	1	2	3	4	5	6	8	10
conduit, straight								
16mm	290	290	290	177	171	167	158	150
20mm	460	460	460	286	278	270	256	244
25mm	800	800	800	514	500	487	463	442
32mm	1400	1400	1400	900	878	857	818	783
conduit, one bend								
16mm	188	177	167	158	150	143	130	120
20mm	303	286	270	256	244	233	213	196
25mm	543	514	487	463	442	422	388	358
32mm	947	900	857	818	783	750	692	643
conduit, two bends								
16mm	177	158	143	130	120	111	97	86
20mm	286	256	233	213	196	182	159	141
25mm	514	463	422	388	358	333	292	260
32mm	900	818	750	692	643	600	529	474

For 38 mm conduit use the 32 mm factor x 1.4. For 50 mm conduit, use the 32 mm factor x 2.6. For 63 mm conduit use the 32 mm factor x 4.2.

Table 4.17 Trunking factors (terms)

Dimensions of trunking (mm x mm)			Factor
37.5	x	50	767
50	x	50	1037
25	x	75	738
37.5	x	75	1146
50	x	75	1555
75	x	75	2371
25	x	100	993
37.5	x	100	1542
50	x	100	2091
75	x	100	3189
100	x	100	4252

To use the tables, the cable factors for all the conductors must be added. The conduit or trunking selected must have a factor (otherwise called 'term') at least as large as this number.

Example 4.9
The following single-core p.v.c. insulated cables are to be run in a conduit 6 m long with a double set: $8 \times 1.4 \times 2.5$ and 2×6 mm². Choose a suitable size.
Consulting {Table 4.14} gives the following cable factors:
16 for 1 mm², 30 for 2.5 mm² and 58 for 6 mm²
Total cable factor is then $(8 \times 16) + (4 \times 30) + (2 \times 58)$
$= 128 + 120 + 116 = 364$
The term "bend" means a right angle bend or a double set.
{Table 4.16} gives a conduit factor for 20 mm conduit 6 m long with a double set as 233, which is less than 364 and thus too small. The next size has a conduit factor of 422 which will be acceptable since it is larger than 364.
The correct conduit size is **25 mm** diameter.

Example 4.10
The first conduit from a distribution board will be straight and 10 m long. It is to enclose 4×10 mm² and 8×4 mm² cables. Calculate a suitable size.
From {Table 4.14}, cable factors are 105 and 43 respectively. Total cable factor
$= (4 \times 105) + (8 \times 43) = 420 + 344 = 764$
From ({Table 4.15}, a 10 m long straight 25 mm conduit has a factor of 442. This is too small, so the next size, with a factor of 783 must be used.
The correct conduit size is **32 mm** diameter.

Example 4.11
A 1.5 m straight length of conduit from a consumer's unit encloses ten 1.5 mm² and four 2.5 mm² solid conductor p.v.c. insulated cables. Calculate a suitable conduit size.
From {Table 4.15} (which is for short straight runs of conduit) total cable factor will be $(10 \times 27) + (4 \times 39) = 426$
Table 4.16 shows that **20 mm** diameter conduit with a factor of 460 will be necessary.

Example 4.12
A length of trunking is to carry eighteen 10 mm², sixteen 6 mm² twelve 4 mm², and ten 2.5 mm² stranded single p.v.c. insulated cables. Calculate a suitable trunking size.
The total cable factor for trunking is calculated with data from {Table 4.14}.
$$18 \times 10 \text{ mm}^2 \text{ at } 36.3 = 18 \times 36.3 = 653.4$$
$$16 \times 6 \text{ mm}^2 \text{ at } 22.9 = 16 \times 22.9 = 366.4$$
$$12 \times 4 \text{ mm}^2 \text{ at } 15.2 = 12 \times 15.2 = 182.4$$
$$10 \times 2.5 \text{ mm}^2 \text{ at } 11.4 = 10 \times 11.4 = \underline{114.0}$$
Total cable factor $= 1316.2$
From the trunking factor {Table 4.17}, two standard trunking sizes have factors slightly greater than the cable factor, and either could be used. They are **50 mm x 75 mm** at 1555, and **37.5 mm x 100 mm** at 1542.

4.6   Conductor and cable identification
4.6.1   *Conduits   ~ [514-02]*
The 'electrical' colour to distinguish conduits from pipelines of other services is orange (BS 1710). Oversheaths for mineral insulated cables are often the same colour, which is also used to identify trunking and switchgear enclosures.
4.6.2   *Identification of fixed wiring conductors* ~ [514-03&-06]
Colour is used to identify the conductors of a wiring system where it is possible to colour the insulation. Where it is not, numbers are used. The requirements for identification of fixed wiring are shown in {Fig 4.21}. There is as yet no requirement to use brown and blue to identify the phase and neutral conductors of fixed wiring, although this applies to flexible cords and cables (*see* {4.6.3}). The colour green on its own is prohibited, although green and

yellow stripes identify the protective conductor. The functional earth conductor for telecommunication circuits is identified by the colour cream.

cables insulated with

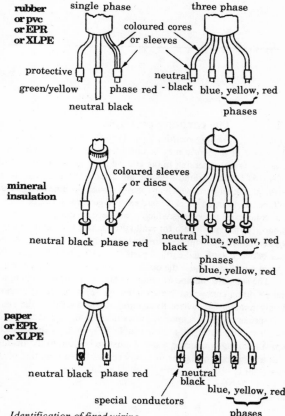

Fig 4.21   Identification of fixed wiring

Some cables comply with HD 324:1977 and have blue insulation on the neutral conductor. This colour does not comply with BS 7671 and if such cables are used, they must be correctly identified at their terminations by the use of black cable markers or black tape.

4.6.3   *Colours for flexible cables and cords* ~ [514-07]

Unlike the cores of fixed cables, which may be identified by sleeves or tapes where they are connected, flexibles must be identified throughout their length. The colour requirements are shown in {Fig 4.22}.

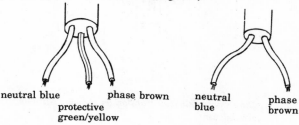

Fig 4.22   Core colours for flexible cables and cords

77

Earthing

5.1 The earthing principle

5.1.1 What is earthing? ~ [130-04, 541 and 542-04]

The whole of the world may be considered as a vast conductor which is at reference (zero) potential. In the UK we refer to this as 'earth' whilst in the USA it is called 'ground'. People are usually more or less in contact with earth, so if other parts which are open to touch become charged at a different voltage from earth a shock hazard exists (see {3.4}). The process of earthing is to connect all these parts which could become charged to the general mass of earth, to provide a path for fault currents and to hold the parts as close as possible to earth potential. In simple theory this will prevent a potential difference between earth and earthed parts, as well as permitting the flow of fault current which will cause the operation of the protective systems.

The standard method of tying the electrical supply system to earth is to make a direct connection between the two. This is usually carried out at the supply transformer, where the neutral conductor (often the star point of a three-phase supply) is connected to earth using an earth electrode or the metal sheath and armouring of a buried cable. {Figure 5.1} shows such a connection. Lightning conductor systems must be bonded to the installation earth with a conductor no larger in cross-sectional area than that of the earthing conductor.

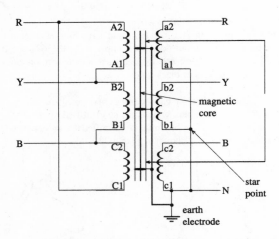

Fig 5.1   Three-phase delta/star transformer showing earthing arrangements

5.1.2 The advantages of earthing

The practice of earthing is widespread, but not all countries in the world use it. There is certainly a high cost involved, so there must be some advantages. In fact there are two. They are:

1. The whole electrical system is tied to the potential of the general mass of earth and cannot 'float' at another potential. For example, we can be fairly certain that the neutral of our supply is at, or near, zero volts (earth potential) and that the phase conductors of our standard supply differ from earth by 240 volts.

2. By connecting earth to metalwork not intended to carry current (an extraneous conductive part or a an exposed conductive part) by using a protective conductor, a path is provided for fault current which can be detected and, if necessary, broken. The path for this fault current is shown in {Fig 5.2}.

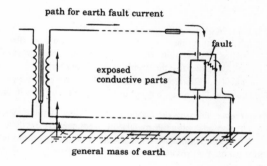

path for earth fault current

fault

exposed
conductive parts

general mass of earth

Fig 5.2   *Path for earth fault current (shown by arrows)*

5.1.3 The disadvantages of earthing

The two important disadvantages are:

1. Cost: the provision of a complete system of protective conductors, earth electrodes, *etc.* is very expensive.

2. Possible safety hazard: It has been argued that complete isolation from earth will prevent shock due to indirect contact because there is no path for the shock current to return to the circuit if the supply earth connection is not made (see {Fig 5.3(a)}). This approach, however, ignores the presence of earth leakage resistance (due to imperfect insulation) and phase-to-earth capacitance (the insulation behaves as a dielectric). In many situations the combined impedance due to insulation resistance and earth capacitive reactance is low enough to allow a significant shock current (see {Fig 5.3(b)}).

5.2 Earthing systems

5.2.1 System classification ~ [312-03, 541-01-01, 542-01-06 and 542-01-09]

The electrical installation does not exist on its own; the supply is part of the overall system. Although Electricity Supply Companies will often provide an earth terminal, they are under no legal obligation to do so. As far as earthing types are concerned, letter classifications are used.

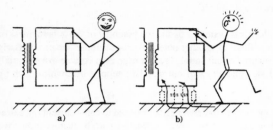

Fig 5.3   *Danger in an unearthed system*
a) apparent safety: no obvious path for shock current
b) actual danger: shock current via stray resistance and capacitance

The first *letter* indicates the type of supply earthing.

T   indicates that one or more points of the supply are directly earthed (for example, the earthed neutral at the transformer).

I   indicates either that the supply system is not earthed at all, or that the earthing includes a deliberately-inserted impedance, the purpose of which is to limit fault current. This method is not used for public supplies in the UK.

The *second letter* indicates the earthing arrangement in the installation.

T   all exposed conductive metalwork is connected directly to earth.

N   all exposed conductive metalwork is connected directly to an earthed supply conductor provided by the Electricity Supply Company.

The *third and fourth letters* indicate the arrangement of the earthed supply conductor system.

S   neutral and earth conductor systems are quite separate.

C   neutral and earth are combined into a single conductor.

A number of possible combinations of earthing systems in common use is indicated in the following subsections.

Protective conductor systems against lightning need to be connected to the installation earthing system to prevent dangerous potential differences. Where a functional earthing system is in use, the protective requirements of the earthing will take precedence over the functional requirements.

5.2.2   **TT systems** ~ [413-02-18 to 413-02-20, 471-08, 542-01-01, 542-01-04 and 542-01-07 to 542-01-09]

This arrangement covers installations not provided with an earth terminal by the Electricity Supply Company. Thus it is the method employed by most (usually rural) installations fed by an overhead supply. Neutral and earth (protective) conductors must be kept quite separate throughout the installation, with the final earth terminal connected to an earth electrode (see {5.5}) by means of an earthing conductor.

Effective earth connection is sometimes difficult. Because of this, socket outlet circuits must be protected by a residual current device (RCD) with an operating current of 30 mA {5.9}. {Fig 5.4} shows the arrangement of a TT earthing system.

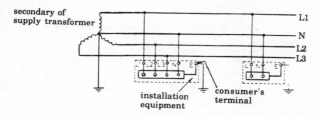

Fig 5.4   TT earthing system

**5.2.3   TN-S system   ~   [413-02-06 to 413-02-17, 542-01-02 and
542-01-07]**

This is probably the most usual earthing system in the UK, with the Electricity Supply Company providing an earth terminal at the incoming mains position. This earth terminal is connected by the supply protective conductor (PE) back to the star point (neutral) of the secondary winding of the supply transformer, which is also connected at that point to an earth electrode. The earth conductor usually takes the form of the armour and sheath (if applicable) of the underground supply cable. The system is shown diagrammatically in {Fig 5.5}.

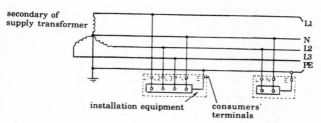

Fig 5.5   TN-S earthing system

**5.2.4   TN-C-S system   ~   [413-02-06 to 413-02-17, 542-01, 542-
03 and 542-07]**

In this system, the installation is TN-S, with separate neutral and protective conductors. The supply, however, uses a common conductor for both the neutral and the earth. This combined earth and neutral system is sometimes called the 'protective and neutral conductor' (PEN) or the 'combined neutral and earth' conductor (CNE). The system, which is shown diagrammatically in {Fig 5.6}, is most usually called the protective multiple earth (PME) system, which will be considered in greater detail in {5.6}.

**5.2.5   TN-C system   ~   [413-02-06 to 413-02-17, 542-01-05 and
542-01-07]**

This installation is unusual, because combined neutral and earth wiring is used in both the supply and within the installation itself. Where used, the installation will usually be the earthed concentric system, which can only be installed under the special conditions listed in {5.7}.

81

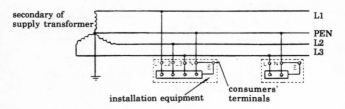

Fig 5.6   TN-C-S earthing system - protective multiple earthing

5.2.6   IT system *[41*3-02-21 to 413-02-26, 542-01-04 and 542-01-07]*
The installation arrangements in the IT system are the same for those of the
TT sytem {5.2.2}. However, the supply earthing is totally different. The IT
system can have an unearthed supply, or one which is not solidly earthed
but is connected to earth through a current limiting impedance.

The total lack of earth in some cases, or the introduction of current
limiting into the earth path, means that the usual methods of protection will
not be effective. For this reason, IT systems are not allowed in the public
supply system in the UK. The method is sometimes used where a supply for
special purposes is taken from a private generator.

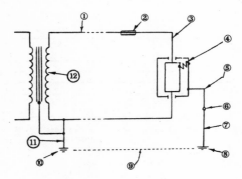

Fig 5.7	*The earth fault loop*
1	*the phase conductor from the transformer to the installation*
2	*the protective device(s) in the installation*
3	*the installation phase conductors from the intake position to the fault*
4	*the fault itself (usually assumed to have zero impedance)*
5	*the protective conductor system*
6	*the main earthing terminal*
7	*the earthing conductor*
8	*the installation earth electrode*
9	*the general mass of earth*
10	*the Supply Company's earth electrode*
11	*the Supply Company's earthing conductor*
12	*the secondary winding of the supply transformer*

5.3　　　Earth fault loop impedance
5.3.1　　*Principle*
The path followed by fault current as the result of a low impedance occurring between the phase conductor and earthed metal is called the earth fault loop. Current is driven through the loop impedance by the supply voltage.

The extent of the earth fault loop for a TT system is shown in {Fig 5.7}, and is made up of the following labelled parts.

For a TN-S system (where the Electricity Supply Company provides an earth terminal), items 8 to 10 are replaced by the PE conductor, which usually takes the form of the armouring (and sheath if there is one) of the underground supply cable.

For a TN-C-S system (protective multiple earthing) items 8 to 11 are replaced by the combined neutral and earth conductor.

For a TN-C system (earthed concentric wiring), items 5 to 11 are replaced by the combined neutral and earth wiring of both the installation and of the supply.

It is readily apparent that the impedance of the loop will probably be a good deal higher for the TT system, where the loop includes the resistance of two earth electrodes as well as an earth path, than for the other methods where the complete loop consists of metallic conductors.

5.3.2　　*The importance of loop impedance*
The earth fault loop impedance can be used with the supply voltage to calculate the earth-fault current.

$$I_F = \frac{U_O}{Z_S}$$

where
I_F = fault current, A
U_O = phase voltage, V
Z_S = loop impedance, Ω

For example, if a 240 V circuit is protected by a 15 A semi-enclosed fuse and has an earth-fault loop impedance of 1.6 Ω, the earth-fault current in the event of a zero impedance earth fault will be

$$I_F = \frac{U_O}{Z_S} = \frac{240}{1.6} \text{ A} = 150 \text{ A}$$

This level of earth-fault current will cause the fuse to operate quickly. From {Fig 3.13} the time taken for the fuse to operate will be about 0.15 s. Any load current in the circuit will be additional to the fault current and will cause the fuse to operate slightly more quickly. However, such load current must not be taken into account when deciding disconnection time, because it is possible that the load may not be connected when the fault occurs.

Note that there is no such thing as a three-phase line/earth fault, although it is possible for three faults to occur on the three lines to earth simultaneously. As far as calculations for fault current are concerned, the voltage to earth for standard UK supplies is always 240 V, for both single-phase and three-phase systems. Thus the Tables of maximum earth-fault loop impedance which will be given in {5.3.4} apply both to single- and to three-phase systems.

5.3.3 *The resistance/impedance relationship*

Resistance, measured in ohms (Ω), is the property of a conductor to limit the flow of current through it when a voltage is applied.

$$I = \frac{U}{R}$$

where I = current, A
U = applied voltage, V
R = circuit resistance, Ω

Thus, a voltage of one volt applied to a one ohm resistance results in a current of one ampere.

When the supply voltage is alternating, a second effect, known as reactance (symbol X) is to be considered. It applies only when the circuit includes inductance and/or capacitance, and its value, measured in ohms, depends on the frequency of the supply as well as on the values of the inductance and/or the capacitance concerned. For almost all installation work the frequency is constant at 50 Hz. Thus, inductive reactance is directly proportional to inductance and capacitive reactance is inversely proportional to capacitance.

$$X_L = 2\pi fL \qquad \text{and} \quad X_C = \frac{1}{2\pi fC}$$

where X_L = inductive reactance (Ω)
X_C = capacitive reactance (Ω)
π = the mathematical constant (3.142)
f = the supply frequency (Hz)
L = circuit inductance (H)
C = circuit capacitance (F)

Resistance (R) and reactance (X_L or X_C) in series add together to produce the circuit impedance (symbol Z), but not in a simple arithmetic manner. Impedance is the effect which limits alternating current in a circuit containing reactance as well as resistance.

$$Z = \frac{U}{I} \qquad \textit{where} \qquad Z = \text{impedance } (\Omega)$$

U = applied voltage (V)
I = current (A)

It follows that a one volt supply connected across a one ohm impedance results in a current of one ampere.

When resistance and reactance are added this is done as if they were at right angles, because the current in a purely reactive circuit is 90° out of phase with that in a purely resistive circuit. The relationships between resistance, reactance and impedance are:

$$Z = \sqrt{(R^2 + X_L^2)} \quad \text{and} \quad Z = \sqrt{(R^2 + X_C^2)}$$

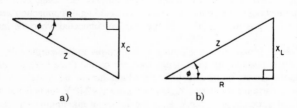

Fig 5.8   *Impedance diagrams   a) resistive and capacitive circuit*
                       b) resistive and inductive circuit

These relationships can be shown in the form of a diagram applying Pythagorus' theorem as shown in {Fig 5.8}. The two diagrams are needed because current lags voltage in the inductive circuit, but leads it in the capacitive. The angle between the resistance R and the impedance Z is called the circuit phase angle, given the symbol ø (Greek 'phi'). If voltage and current are both sinusoidal, the cosine of this angle, cos ø, is the circuit power factor, which is said to be lagging for the inductive circuit, and leading for the capacitive.

In practice, all circuits have some inductance and some capacitance associated with them. However, the inductance of cables only becomes significant when they have a cross-sectional area of 25 mm^2 and greater. Remember that the higher the earth fault loop impedance the smaller the fault current will be. Thus, if simple arithmetic is used to add resistance and reactance, and the resulting impedance is low enough to open the protective device quickly enough, the circuit will be safe. This is because the Pythagorean addition will always give lower values of impedance than simple addition.

For example, if resistance is $2 \ \Omega$ and reactance $1 \ \Omega$, simple arithmetic addition gives

$$Z = R + X = 2 + 1 = 3 \ \Omega$$

and correct addition gives

$$Z = \sqrt{(R^2 + X^2)}$$
$$= \sqrt{(2^2 + 1^2)} = \sqrt{5} = 2.24 \ \Omega$$

If $3 \ \Omega$ is acceptable, $2.24 \ \Omega$ will allow a larger fault current to flow which will operate the protective device more quickly and is thus even more acceptable.

5.3.4   *Earth-fault loop impedance values* ~ [413-02-07 to 413-02-14 and 471-08]

The over-riding requirement is that sufficient fault current must flow in the event of an earth fault to ensure that the protective device cuts off the supply before dangerous shock can occur. For normal 240 V systems, there are two levels of maximum disconnection time. These are:

For socket outlet circuits where equipment could be tightly grasped: *0.4 s*

For fixed equipment where contact is unlikely to be so good: *5 s*

The maximum disconnection time of 5 s also applies to feeders and submains.

It must be appreciated that the longest disconnection times for protective devices, leading to the longest shock times and the greatest danger, will be associated with the lowest levels of fault current, and not, as is commonly believed, the highest levels.

Where the voltage is other than 240 V, [Table 41A] gives a range of disconnection times for socket outlet circuits, of which the lowest is 0.1 s for voltages exceeding 400 V.

In general, the requirement is that if a fault of negligible impedance occurs between a phase and earth, the earth-fault loop impedance must not be greater than the value calculated from:

$$Z_s < \frac{U_O}{Ia}$$

where Z_S = the earth fault loop impedance (Ω)

           U_O = the system voltage to earth (V)

           I_a = the current causing automatic disconnection (operation of the protective device) in the required time [A]).

The earth fault loop values shown in [Tables 5.1, 5.2 and 5.4] depend on the supply voltage and assume, as shown in the Tables, a value of 240 V. Whilst it would appear that 240 V is likely to be the value of the supply voltage in Great Britain for the foreseeable future, it is not impossible that different values may apply. In such a case, the tabulated value for earth fault loop impedance should be modified using the formula:-

$$Z_s = Z_t \times \frac{U}{U_{240}}$$

where Z_S is the earth fault loop impedance required for safety

                  Z_t is the tabulated value of earth fault loop impedance

                  U is the actual supply voltage

                  U_{240} is the supply voltage assumed in the Table.

As an alternative to this calculation, a whole series of maximum values of earth fault loop impedance is given in {Table 5.1} (from [Table 41B]) for disconnection within 0.4 s. The reader should not think that these values are produced in some mysterious way — all are easily verified using the characteristic curves {Figs 3.13 to 3.19}.

For example, consider a 20 A HBC fuse to BS88 used in a 240 V system. The fuse characteristic is shown in {Fig 3.15}, and indicates that disconnection in 0.4 s requires a current of about 130 A. It is difficult (if not impossible) to be precise about this value of current, because it is between the 100 A and 150 A current graduations.

Using these values,

$$Z_s = \frac{U_O}{I_a} = \frac{240}{130} \ \Omega = 1.84 \ \Omega$$

Reference to {Table 5.1} shows that the stated value is 1.8 Ω, the discrepancy being due to the difficulty in reading the current with accuracy. {Tables 5.1 and 5.2} give maximum earth-fault loop impedance values for fuses and for miniature circuit breakers to give a minimum disconnection time of 0.4 s in the event of a zero impedance fault from phase to earth.

The reason for the inclusion of fixed equipment as well as distribution circuits in {Table 5.2} will become apparent later in this sub-section.

The severity of the electric shock received when there is a phase to earth fault (indirect contact) depends entirely on the impedance of the circuit protective conductor. We saw in {3.4.3} and {Fig 3.8} how the volt drop across the protective conductor is applied to the person receiving the shock. Since this volt drop is equal to fault current times protective conductor impedance, if the protective conductor has a lower impedance the shock voltage will be less. Thus it can be sustained for a longer period without extreme danger.

Table 5.1 Maximum earth-fault loop impedance values for 240 V socket outlet circuits protected by fuses
(from [Table 41B1] of BS 7671: 1992)

Fuse rating (A)	Maximum earth-fault loop impedance (Ω)		
	Cartridge BS 88	Cartridge BS 1361	Semi- enclosed BS3036
5	-	10.9	10.0
6	8.89	-	-
10	5.33	-	-
15	-	3.43	2.67
20	1.85	1.78	1.85
30	-	1.20	1.14
32	1.09	-	-
40	0.86	-	-
45	-	0.60	0.62

Table 5.2 Maximum earth-fault loop impedance values for 240 V socket outlet circuits, fixed equipment and distribution circuits protected by miniature circuit breakers
(from [Table 41B2] of BS 7671: 1992)

Device rating (A)	Maximum earth-fault loop impedance (Ω)				
	MCB type 1	MCB type 2	MCB type 3 and type C	MCB type B	MCB type D
5	12.00	6.86	4.80	—	2.40
6	10.00	5.71	4.00	8.00	2.00
10	6.00	3.43	2.40	4.80	1.20
15	4.00	2.29	1.60	—	0.80
16	3.75	2.14	1.50	3.00	0.75
20	3.0	1.71	1.20	2.40	0.60
25	2.40	1.37	0.96	1.92	0.48
30	2.00	1.14	0.80	—	0.40
32	1.88	1.07	0.75	1.50	0.38
40	1.50	0.86	0.60	1.20	0.30

Table 5.3 Maximum impedance of circuit protective conductors to allow 5 s disconnection time for socket outlets
(from [Table 41C] of BS 7671: 1992)

Rating (A)	Maximum impedance of circuit protective conductor (Ω)							
	Fuse BS 88	Fuse BS 1361	Fuse BS 3036	MCB type 1	MCB type 2	MCB types 3&C	MCB type B	MCB type D
5	—	3.25	3.25	2.50	1.43	1.00	—	0.50
6	2.48	—	—	2.08	1.19	0.83	1.67	0.42
10	1.48	—	—	1.25	0.71	0.50	1.00	0.25
15	—	0.96	0.96	0.83	0.48	0.33	—	—
16	0.83	—	—	0.78	0.45	0.31	0.63	0.16
20	0.55	0.55	0.63	0.63	0.36	0.25	0.50	0.12
25	0.43	—	—	—	—	—	—	0.10
30	—	0.36	0.43	0.42	0.24	0.17	—	—
32	0.34	—	—	0.39	0.22	0.16	0.31	0.08
40	0.26	—	—	0.31	0.18	0.13	0.25	0.06
45	—	0.18	0.24	0.28	0.16	0.11	0.22	0.06

Socket outlet circuits can therefore have a disconnection time of up to 5 s provided that the circuit protective conductor impedances are no higher than shown in {Table 5.3} for various types of protection.

The reasoning behind this set of requirements becomes clearer if we take an example. {Table 5.3} shows that a 40 A cartridge fuse to BS 88 must have an associated protective conductor impedance of no more than 0.29 Ω if it is to comply. Now look at the time/current characteristic for the fuse {Fig 3.15} from which we can see that the current for operation in 5 s is about 170 A. The maximum volt drop across the conductor (the shock voltage) is thus 170 x 0.29 or 49.3 V. Application of the same reasoning to all the figures gives shock voltages of less than 50 V. This limitation on the impedance of the CPC is of particular importance in TT systems where it is likely that the resistance of the earth electrode to the general mass of earth will be high.

The breaking time of 5 s also applies to fixed equipment, so the earth-fault loop impedance values can be higher for these circuits, as well as for distribution circuts. For fuses, the maximum values of earth-fault loop impedance for fixed equipment are given in {Table 5.4}.

No separate values are given for miniature circuit breakers. Examination of the time/current characteristics {Figs 3.16 to 3.19} will reveal that there is no change at all in the current causing operation between 0.4 s and 5 s in all cases except the Type 1. Here, the vertical characteristic breaks off at 4 s, but this makes little difference to the protection. In this case, the values given in {Table 5.2} can be used for fixed equipment as well as for socket outlet circuits. An alternative is to calculate the loop impedance as described above.

Table 5.4 Maximum earth-fault loop impedance values for 240 V fixed equipment and distribution circuits protected by fuses
(from [Table 41D] of BS 7671: 1992)

Device rating (A)	*Maximum earth-fault loop impedance (Ω)*		
	Cartridge BS 88	*Cartridge BS 1361*	*Semi-enclosed BS3036*
5	—	17.1	18.5
6	14.1	—	—
10	7.74	—	—
15	—	5.22	5.58
16	4.36	—	—
20	3.04	2.93	4.00
30	—	1.92	2.76
32	1.92	—	—
40	1.41	—	—
45	—	1.00	1.66
50	1.09	—	—

5.3.5   ***Protective conductor impedance*** ~ [413-02-12 to 413-02-14, 471-08-01 to 471-08-05 and 544]

It has been shown in the previous sub-section how a low-impedance protective conductor will provide safety from shock in the event of a fault to earth. This method can only be used where it is certain that the shock victim can never be in contact with conducting material at a different potential from that of the earthed system in the zone he occupies. Thus, all associated exposed or extraneous parts must be within the equipotential zone (see {5.4}). When overcurrent protective devices are used as protection from electric shock, the protective conductor must be in the same wiring system as, or in close proximity to, the live conductors. This is intended to ensure that the protective conductor is unlikely to be damaged in an accident without the live conductors also being cut.

{Figure 5.9} shows a method of measuring the resistance of the protective conductor, using a line conductor as a return and taking into account the different cross-sectional areas of the phase and the protective conductors.

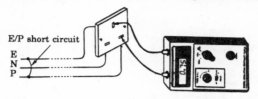

E/P short circuit

E
N
P

Fig 5.9   Measurement of protective conductor resistance

Taking the cross-sectional area of the protective conductor as A_p and that of the line (phase or neutral) conductor as A_L , then

$$R_p = \text{resistance reading} \quad \times \quad \frac{A_L}{A_L + A_p}$$

For example, consider a reading of 0.72 Ω obtained when measuring a circuit in the way described and having 2.5 mm^2 line conductors and a 1.5 mm^2 protective conductor. The resistance of the protective conductor is calculated from:

$$R_p = R \times \frac{A_L}{A_L + A_p} = \frac{0.72 \times 2.5}{2.5 + 1.5} \ \Omega$$
$$= \frac{0.72 \times 2.5}{4.0} \ \Omega = 0.45 \ \Omega$$

5.3.6   *Maximum circuit conductor length* [543-01]

The complete earth-fault loop path is made up of a large number of parts as shown in {Fig 5.7}, many of which are external to the installation and outside the control of the installer. These external parts make up the external loop impedance (Z_E). The rest of the earth-fault loop impedance of the installation consists of the impedance of the phase and protective conductors from the intake position to the point at which the loop impedance is required.

Since an earth fault may occur at the point farthest from the intake position, where the impedance of the circuit conductors will be at their highest value, this is the point which must be considered when measuring or calculating the earth-fault loop impedance for the installation. Measurement of the impedance will be considered in {8.6.2}. Provided that the external fault loop impedance value for the installation is known, total impedance can be calculated by adding the external impedance to that of the installation conductors to the point concerned. The combined resistance of the phase and protective conductors is known as $R_1 + R_2$. The same term is sometimes used for the combined resistance of neutral and protective conductors (*see* {8.4.1}). In the vast majority of cases phase and neutral conductors have the same cross-sectional area and hence the same resistance.

For the majority of installations, these conductors will be too small for their reactance to have any effect (below 25 mm^2 cross-sectional area reactance is very small), so only their resistances will be of importance. This can be measured by the method indicated in {Fig 5.9}, remembering that this time we are interested in the combined resistance of phase and protective conductors, or can be calculated if we measure the cable length and can find data concerning the resistance of various standard cables. These data are given here as {Table 5.5}.

Table 5.5 Resistance per metre of copper conductors for calculation of $R_1 + R_2$	
Conductor cross-sectional area(mm²)	*Resistance per metre run (mΩ/m)*
1.0	18.10
1.5	12.10
2.5	7.41
4.0	4.61
6.0	3.08
10.0	1.83
16.0	1.15
25.0	0.727

NOTE that to allow for the increase in resistance with increased temperature under fault conditions the values of {Table 5.5} **MUST BE MULTIPLIED BY 1.2 FOR PVC INSULATED CABLES** (*see* {Table 8.7}).

The resistance values given in {Table 5.5} are for cables at normal temperatures. Under fault conditions the high fault current will cause the temperature of the cables to rise and result in an increase in resistance. To allow for this changed resistance, the values in {Table 5.5} must be multiplied by the appropriate correction factor from Table {8.7}. It should be mentioned that the practice which has been adopted here of adding impedance and resistance values arithmetically is not strictly correct. Phasor addition is the only perfectly correct method since the phase angle associated with resistance is likely to be different from those associated with impedance, and in addition impedance phase angles will differ from one another. However, if the phase angles are similar, and this will be so in the vast majority of cases where electrical installations are concerned, the error will be acceptably small.

It is often assumed that higher conductor temperatures are associated with the higher levels of fault current. In most cases this is untrue. A lower fault level will result in a longer period of time before the protective device operates to clear it, and this often results in higher conductor temperature.

Example 5.1

A 7 kW shower heater is to be installed in a house fed with a 240 V TN-S supply system with an external loop impedance (Z_E) of 0.8 Ω. The heater is to be fed from a 32 A BS 88 cartridge fuse. Calculate a suitable size for the p.v.c. insulated and sheathed cable to be used and determine the maximum possible length of run for this cable. It may be assumed that the cable will not be subject to any correction factors and is clipped direct to a heat conducting surface throughout its run.

First calculate the circuit current.

$$I = \frac{P}{U} = \frac{7000}{240} \text{ A} = 29.2 \text{ A}$$

Next, select the cable size from {Table 4.7} (which is based on [Table 4D2A]) from which we can see that 4 mm² cable of this kind clipped direct has a rating of 36 A. The 2.5 mm² rating of 27 A is not large enough. We shall assume that a 2.5 mm² protective conductor is included within the sheath of the cable.

Now we must find the maximum acceptable earth-fault loop impedance for the circuit. Since this is a shower, a maximum disconnection time of 0.4 s will apply, so we need to consult {Table 5.1}, which gives a maximum loop impedance of 1.09 Ω for this situation. Since the external loop impedance is 0.8 Ω, we can calculate the maximum resistance of the cable.

R_{cable} = max. loop impedance - external impedance

$$= 1.09 - 0.8 \ \Omega \ = \ 0.29 \, \Omega$$

This assumes that resistance and impedance phase angles are identical, which is not strictly the case. However, the difference is unimportant.

The phase conductor (4 mm^2) has a resistance of 4.6 mΩ/m and the 2.5 mm^2 protective conductor 7.4 mΩ/m, so the combined resistance per metre is 4.6 + 7.4 = 12.0 mΩ/m. The cable will get hot under fault conditions, so we must apply the multiplier of 1.2.

Effective resistance of cable per metre = 12.0 x 1.2 mΩ/m

$$= 14.4 \ \text{m}\Omega/\text{m}$$

The maximum length of run in metres is thus the number of times 14.4 mΩ will divide into 0.29 Ω.

Maximum length of run $= \dfrac{0.29 \ \text{x} \ 1000}{14.4} \ \text{m} \ = 20.1 \ \text{m}$

This is not quite the end of our calculation, because we must check that this length of run will not result in an excessive volt drop. From {Table 4.7} based on [Table 4D2B] a 4 mm^2 cable of this kind gives a volt drop of 11 mV/A/m.

Volt drop = 11 x circuit current (A) x length of run (m) divided by 1000

Volt drop $= \dfrac{11 \ \text{x} \ 29.2 \ \text{x} \ 20.1}{1000} \ \text{V} \ = \ 6.46 \ \text{V}$

Since the permissible volt drop is 4% of 240 V or 9.6 V, this length of run is acceptable.

5.4 Protective conductors

5.4.1 *Earthing conductors* ~ [514-13-01, 542-01-08 and 542-03]

The earthing conductor is commonly called the earthing lead. It joins the installation earthing terminal to the earth electrode or to the earth terminal provided by the Electricity Supply Company. It is a vital link in the protective system, so care must be taken to see that its integrity will be preserved at all times. Aluminium conductors and cables may now be used for earthing and bonding, but great care must be taken when doing so to ensure that there will be no problems with corrosion or with electrolytic action where they come into contact with other metals.

Where the final connection to the earth electrode or earthing terminal is made there must be a clear and permanent label SAFETY ELECTRICAL CONNECTION - DO NOT REMOVE (*see* {Fig 5.17}). Where a buried earthing conductor is not protected against mechanical damage but is protected against corrosion by a sheath, its minimum size must be 16 mm^2, whether made of copper or coated steel. If it has no corrosion protection, minimum sizes for mechanically unprotected earthing conductors are 25 mm^2 for copper and 50 mm^2 for coated steel.

If not protected against corrosion the latter sizes again apply, whether protected from mechanical damage or not.

Earthing conductors, as well as protective and bonding conductors, must be protected against corrosion. Probably the most common type of corrosion is electrolytic, which is an electro-chemical effect between two different metals when a current passes between them whilst they are in contact with each other and with a weak acid. The acid is likely to be any moisture which has become contaminated with chemicals carried in the air or in the ground. The effect is small on ac supplies because any metal removed whilst current flows in one direction is replaced as it reverses in the next half cycle. For dc systems, however, it will be necessary to ensure that the system remains perfectly dry (a very difficult task) or to use the 'sacrificial anode' principle.

A main earth terminal or bar must be provided for each installation to collect and connect together all protective and bonding conductors. It must be possible to disconnect the earthing conductor from this terminal for test purposes, but only by the use of a tool. This requirement is intended to prevent unauthorised or unknowing removal of protection.

5.4.2 *Protective conductor types* ~ [543-02 and 543-03]

The circuit protective conductor (increasingly called the 'c.p.c.') is a system of conductors joining together all exposed conductive parts and connecting them to the main earthing terminal. Strictly speaking, the term includes the earthing conductor as well as the equipotential bonding conductors.

The circuit protective conductor can take many forms, such as:

1. a separate conductor which must be green/yellow insulated if of less than 10 mm^2 cross-sectional area.
2. a conductor included in a sheathed cable with other conductors
3. the metal sheath and/or armouring of a cable
4. conducting cable enclosures such as conduit or trunking
5. exposed conductive parts, such as the conducting cases of equipment

This list is by no means exhaustive and there may be many other items forming parts of the circuit protective conductor as indicated in {Fig 5.10}. Note that gas or oil pipes must not be used for the purpose, because of the possible future change to plastic (non-conducting) pipes.

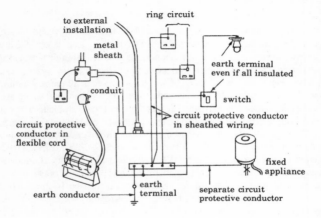

Fig 5.10   Some types of circuit protective conductor

It is, of course, very important that the protective conductor remains effective throughout the life of the installation. Thus, great care is needed to ensure that steel conduit used for the purpose is tightly jointed and unlikely to corrode. The difficulty of ensuring this point is leading to the increasing use of a c.p.c. run inside the conduit with the phase conductors. Such a c.p.c. will, of course, always be necessary where plastic conduits are used. Where an accessory is connected to a system (for example, by means of a socket outlet) which uses conduit as its c.p.c., the appliance (or socket outlet) earthing terminal must be connected by a separate conductor to the earth terminal of the conduit box (see {Fig 5.11}). This connection will ensure that the accessory remains properly earthed even if the screws holding it into the box become loose, damaged or corroded.

A separate protective conductor will be needed where flexible conduit is used, since this type of conduit cannot be relied upon to maintain a low resistance conducting path (*see* {Fig 5.12}).

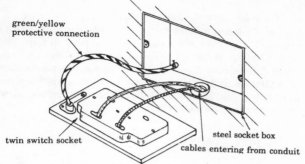

green/yellow
protective connection

steel socket box

twin switch socket

cables entering from conduit

Fig 5.11   Protective connection for socket outlet in conduit system

5.4.3   *Bonding conductors* ~ [130-04-04, 413-02-27 and 413-02-28, 514-13-01, 541, 546 and 547]

The purpose of the protective conductors is to provide a path for earth fault current so that the protective device will operate to remove dangerous potential differences, which are unavoidable under fault conditions, before a dangerous shock can be delivered. Equipotential bonding serves the purpose of ensuring that the earthed metalwork (exposed conductive parts) of the installation is connected to other metalwork (extraneous conductive parts) to ensure that no dangerous potential differences can occur. The resistance of such a bonding conductor must be low enough to ensure that its volt drop when carrying the operating current of the protective device never exceeds 50 V.

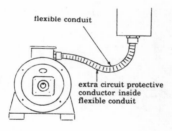

flexible conduit

extra circuit protective
conductor inside
flexible conduit

Fig 5.12   Separate additional protective conductor with flexible conduit

Thus $R < \dfrac{50}{I_a}$.

where   R   is the resistance of the bonding conductor
           I_a is the operating current of the protective device.

Two types of equipotential bonding conductor are specified.

1   *Main equipotential bonding conductors*

These conductors connect together the installation earthing system and the metalwork of other services such as gas and water. This bonding of service pipes must be effected as close as possible to their point of entry to the building, as shown in {Fig 5.13}. Metallic sheaths of telecommunication cables must be bonded, but the consent of the owner of the cable must be obtained before doing so. The minimum size of bonding conductors is related to the size of the main supply conductors (the tails) and is given in {Table 5.6}.

2 *Supplementary bonding conductors*

These conductors connect together extraneous conductive parts — that is, metalwork which is not associated with the electrical installation but which may provide a conducting path giving rise to shock. The object is to ensure that potential differences in excess of 50 V between accessible metalwork cannot occur; this means that the resistance of the bonding conductors must be low (*see* {Table 5.7}). {Figure 5.14} shows some of the extraneous metalwork in a bathroom which must be bonded.

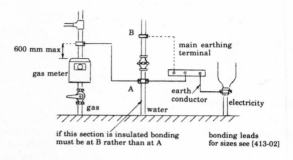

Fig 5.13   *Main bonding connections*

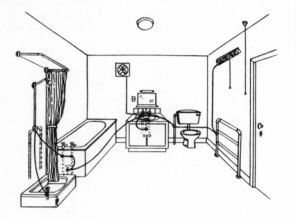

Fig 5.14   *Supplementary bonding in a bathroom*

The cross-sectional areas required for supplementary bonding conductors are shown in {Table 5.6}. Where connections are between extraneous parts only, the conductors may be 2.5 mm^2 if mechanically protected or 4 mm^2 if not protected. If the circuit protective conductor is larger than 10 mm^2, the supplementary bonding conductor must have have at least half this cross-sectional area. Supplementary bonding conductors of less than 16 mm^2 cross sectional area must not be aluminium. {Fig 5.15} shows the application of a supplementary bonding conductor to prevent the severe shock which could otherwise occur between the live case of a faulty electric kettle and an adjacent water tap.

There will sometimes be doubt if a particular piece of metalwork should be bonded. The answer must always be that bonding will be necessary if there is a danger of severe shock when contact is made between a live system and the metalwork in question. Thus if the resistance between the metalwork and the general mass of earth is low enough to permit the passage of a dangerous shock current, then the metalwork must be bonded.

The question can be resolved by measuring the resistance (R_x) from the metalwork concerned to the main earthing terminal. Using this value in the formula:

$$I_b = \frac{U_O}{R_p + R_x}$$

will allow calculation of the maximum current likely to pass through the human body *where:*

I_b is the shock current through the body (A)

U_O is the voltage of the supply (V)

R_p is the resistance of the human body (Ω) and

R_x is the measured resistance from the metalwork concerned to the main earthing terminal (Ω).

The resistance of the human body, R_p can in most cases be taken as $1000\ \Omega$ although $200\ \Omega$ would be a safer value if the metalwork in question can be touched by a person in a bath. Although no hard and fast rules are possible for the value of a safe shock current, I_b, it is probable that 10 mA is seldom likely to prove fatal. Using this value with 240 V for the supply voltage, U_O, and $1000\ \Omega$ as the human body resistance, R_p, the minimum safe value of R_p calculates to 23 kΩ. If the safer values of 5 mA for I_b and $200\ \Omega$ for Rp are used, the value of R_x would be 47.8 kΩ for a 240 V supply.

To sum up, when in doubt about the need to bond metalwork, measure its resistance to the main earthing terminal. If this value is 50 kΩ or greater, no bonding is necessary. In a situation where a person is not wet, bonding could be ignored where the resistance to the main earthing terminal is as low as 25 kΩ. To reduce the possibility of bonding conductors being disconnected by those who do not appreciate their importance, every bonding connection should be provided with a label like that shown in Fig. 5.17.

Table 5.6 Supplementary bonding conductor sizes

Circuit protective conductor size	Supplementary bonding conductor size	
	not protected	mechanically protected
1.0 mm^2	4.0 mm^2	2.5 mm^2
1.5 mm^2	4.0 mm^2	2.5 mm^2
2.5 mm^2	4.0 mm^2	2.5 mm^2
4.0 mm^2	4.0 mm^2	2.5 mm^2
6.0 mm^2	4.0 mm^2	4.0 mm^2
10. mm^2	6.0 mm^2	6.0 mm^2

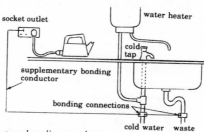

Fig 5.15   *Supplementary bonding conductor*

5.4.4   *Protective conductor cross-section assessment* ~   [543-01-01,543-01-02 and 543-01-04]

A fault current will flow when an earth fault occurs. If this current is large enough to operate the protective device quickly, there is little danger of the protective conductor and the exposed conductive parts it connects to earth being at a high potential to earth for long enough for a dangerous shock to occur. The factors determining the fault current are the supply voltage and the earth-fault loop impedance *(see* {5.3}).

The earth fault results in the protective conductors being connected in series across the supply voltage {Fig 5.16}. The voltage above earth of the earthed metalwork (the voltage of the junction between the protective and phase conductors) at this time may become dangerously high, even in an installation complying with the Regulations. The people using the installation will be protected by the ability of the fuse or circuit breaker in a properly designed installation to cut off the supply before dangerous shock damage results.

Remember that lower fault levels result in a longer time before operation of the protective device. Since the cross-sectional area of the protective conductor will usually be less than that of live conductors, its temperature, and hence its resistance, will become higher during the fault, so that the shock voltage will be a higher proportion of the supply potential *(see* {Fig 5.16}).

{Fig 5.16} shows the circuit arrangements, with some typical phase- and protective-conductor resistances. In this case, a shock voltage of 140 V will be applied to a person in contact with earthed metal and with the general mass of earth. Thus, the supply must be removed very quickly. The actual voltage of the shock depends directly on the relationship between the phase conductor resistance and the protective conductor resistance. If the two are equal, exactly half the supply voltage will appear as the shock voltage.

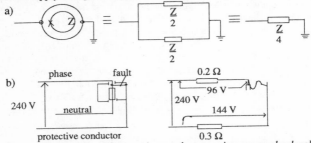

Fig 5.16   The effect of protective conductor resistance on shock voltage
a) effective resistance of a ring circuit protective conductor
b) potential differences across healthy protective conductor in the event of
an earth fault

For socket outlet circuits, where the shock danger is highest, the maximum protective conductor resistance values of {Table 5.3} will ensure that the shock voltage never exceeds the safe value of 50 V. If the circuit concerned is in the form of a ring, one quarter of the resistance of the complete protective conductor round the ring must not be greater than the {Table 5.3} figure. The reason for this is shown in {Fig 5.16(a)}. This assumes that the fault will occur exactly at the mid point of the ring. If it happens at any other point, effective protective conductor resistance is lower, and safer, than one quarter of the total ring resistance.

{Table 5.7} allows selection (rather than calculation) of sizes for earthing and bonding conductors. The rules applying to selection are:
For phase conductors up to 16 mm^2, the protective conductor has the same size as the phase conductor
For phase conductors from 16 mm^2 to 35 mm^2, the protective conductor must be 16 mm^2
For phase conductors over 35 mm^2, the protective conductor must have at least half the c.s.a. of the phase conductor.

Table 5.7 Main earthing and main equipotential bonding conductor sizes for TN-S and TN-C-S supplies

Phase conductor (or neutral for PME supplies) csa mm^2	Earthing conductor (not buried or protected against mechanical damage) csa mm^2	Main equipotential bonding conductor csa mm^2	Main equipotential bonding conductor for PME supplies csa mm^2
4	4	6	10
6	6	6	10
10	10	6	10
16	16	10	10
25	16	10	10
35	16	10	10
50	25	16	16
70	35	16	25

The minimum cross-sectional area of a separate CPC is 2.5 mm^2 if mechanically protected and 4mm^2 if not.

Note that Regional Electricity Companies may require a minimum size of earthing conductor of 16 mm^2 at the origin of the installation. Always consult them before designing an installation.

5.4.5 *Protective conductor cross-section calculation* ~ [543-01-03]
The c.s.a. of the circuit protective conductor (c.p.c.) is of great importance since the level of possible shock in the event of a fault depends on it (as seen in {5.4.4}). Safety could always be assured if we assessed the size using {Table 5.7} as a basis. However, this would result in a more expensive installation than necessary because we would often use protective conductors which are larger than those found to be acceptable by calculation. For example, twin with cpc insulated and sheathed cables larger than 1 mm^2 would be ruled out because in all other sizes the cpc is smaller than required by {Table 5.7}.

In very many cases, calculation of the cpc size will show that a smaller size than that detailed in {5.4.4} is perfectly adequate. The formula to be used is:

$$S = \frac{\sqrt{(I_a^2 t)}}{k} \quad mm^2 \quad where$$

S   is the minimum protective conductor cross-sectional area (mm^2)
I_a is the fault current (A)
t   is the opening time of the protective device (s)
k   is a factor depending on the conductor material and insulation, and the initial and maximum insulation temperatures.

This is the same formula as in {3.7.3}, the adiabatic equation, but with a change in the subject. To use it, we need to have three pieces of information, I_a, t and k.

1) To find I_a
Since $I_a = \dfrac{U_O}{Z_S}$ we need values for U_O and for Z_S.

U_O is simply the supply voltage, which in most cases will be 240V.
Z_S is the earth-fault loop impedance assuming that the fault has zero impedance.

Since we must assume that we are at the design stage, we cannot measure the loop impedance and must calculate it by adding the loop impedance external to the installation (Z_E) to the resistance of the conductors to the furthest point in the circuit concerned. This technique was used in {5.3.6}.

Thus, $Z_S = Z_E + R_1 + R_2$ where R_1 and R_2 are the resistances of the phase and protective conductors respectively from {Table 5.5}.

2) To find **t**

We can find *t* from the time/current characteristics of {Figs 3.13 to 3.19} using the value of I $_a$ already calculated above. For example, if the protective device is a 20 A miniature circuit breaker type 1 and the fault current is 1000 A, we shall need to consult {Fig 3.16}, when we can read off that operation will be in 0.01 s (10 ms). (It is of interest here to notice that if the fault current had been 80 A the opening time could have been anything from 0.04 s to 20 s, so the circuit would not have complied with the required opening times).

3) To find **k**

k is a constant, which we cannot calculate but must obtain from a suitable table of values. Some values of *k* for typical protective conductors are given in {Table 5.8}.

It is worth pointing out here that correctly installed steel conduit and trunking will always meet the requirements of the Regulations in terms of protective conductor impedance.

Although appearing a little complicated, calculation of acceptable protective conductor size is worth the trouble because it often allows smaller sizes than those shown in {Table 5.7}.

Table 5.8 Values of *k* for protective conductors
(from {Tables 54B, 54C, 54E and 54F} of BS 7671: 1992)

Nature of protective conductor	Initial temp (^oC)	Final temp. (^oC)	Conductor material	k
p.v.c. insulated, not in cable or bunched	30	160	copper	143
	30	160	aluminium	95
	30	160	steel	52
p.v.c. insulated, in cable or bunched	70	160	copper	115
	70	160	aluminium	76
steel conduit or trunking	50	160	steel	47
bare conductor	30	200	copper	159
	30	200	aluminium	105
	30	200	steel	58

Example 5.2

A load takes 30 A from a 240 V single phase supply and is protected by a 32 A HBC fuse to BS 88. The wiring consists of 4 mm^2 single core p.v.c. insulated cables run in trunking, the length of run being 18 m. The earth-fault loop impedance external to the installation is assessed as 0.7 Ω. Calculate the cross-sectional area of a suitable p.v.c. sheathed protective conductor.

This is one of those cases where we need to make an assumption of the answer to the problem before we can solve it. Assume that a 2.5 mm^2 protective conductor will be acceptable and calculate the combined resistance of the phase and protective conductors from the origin of the installation to the end of the circuit. From {Table 5.5}, 2.5 mm^2 cable has a resistance of 7.4 mΩ/m and 4 mm^2 a resistance of 4.6 mΩ/m. Both values must be multipled by 1.2 to allow for increased resistance as temperature rises due to fault current.

Thus, $R_1 + R_2 = \dfrac{(7.4 + 4.6) \times 1.2 \times 18}{1000}$ Ω $= \dfrac{12.0 \times 1.2 \times 18}{1000}$ Ω $= 0.26$ Ω

This conductor resistance must be added to external loop impedance to give the total

earth-fault loop impedance.
$$Z_s = Z_E + R_1 + R_2 = 0.7 + 0.26 \, \Omega = 0.96 \, \Omega$$
We can now calculate the fault current:

$$I_a = \frac{U_o}{Z_s} = \frac{250}{0.96} \, A = 240 \, A$$

Next we need to find the operating time for a 32 A BS 88 fuse carrying 250 A. Examination of {Fig 3.15} shows that operation will take place after 0.2 s.

Finally, we need a value for k. From {Table 5.8} we can read this off as 115, because the protective conductor will be bunched with others in the trunking.

We now have values for I_a, t and k so we can calculate conductor size.

$$S = \frac{\sqrt{(I_a^2 t)}}{k} = \frac{\sqrt{(250^2 \times 0.2)}}{115} \, mm^2 = 0.97 \, mm^2$$

This result suggests that a 1.0 mm^2 protective conductor will suffice. However, it may be dangerous to make this assumption because the whole calculation has been based on the resistance of a 2.5 mm^2 conductor. Let us start again assuming a 1.5 mm^2 protective conductor and work the whole thing through again.

The new size protective conductor has a resistance of 18.1 mΩ/m, *see*{Table5.5}, and with the 4 mm^2 phase conductor gives a total conductor resistance, allowing for increased temperature, of 0.491 Ω. When added to external loop impedance this gives a total earth-fault loop impedance of 1.191 Ω and a fault current at 240 V of 202 A. From {Fig 3.15} operating time will be 0.6 s. The value of k will be unchanged at 115.

$$S = \frac{\sqrt{(I_a^2 t)}}{k} = \frac{\sqrt{(202^2 \times 0.6)}}{115} \, mm^2 = 1.36 \, mm^2$$

Thus, a 1.5 mm^2 protective conductor can be used in this case. Note that if the size had been assessed rather than calculated, the required size would be 4 mm^2, two sizes larger. A point to notice here is that the disconnection time with a 1.5 mm^2 protective conductor is 0.6 s, which is too long for socket outlet circuits (0.4 s maximum).

Example 5.3

A 240 V, 30 A ring circuit for socket outlets is 45 m long and is to be wired in 2.5 mm^2 flat twin p.v.c. insulated and sheathed cable incorporating a 1.5 mm^2 cpc. The circuit is to be protected by a semi-enclosed (rewirable) fuse to BS 3036, and the earth-fault loop impedance external to the installation has been ascertained to be 0.3 Ω. Verify that the 1.5 mm^2 cpc enclosed in the sheath is adequate.

First use [Table 5.5] to find the resistance of the phase and cpc conductors. These are 7.4 mΩ/m and 12.1 mΩ/m respectively, so for a 45 m length and allowing for the resistance increase with temperature factor of 1.2.

$$R_1 + R_2 = \frac{(7.4 + 12.1) \times 1.2 \times 45}{1000} \, \Omega = \frac{19.5 \times 1.2 \times 45}{1000} \, \Omega = 1.05 \, \Omega$$
$$Z_s = Z_E + \frac{R_1 + R_2}{4} = 0.3 + \frac{1.05}{4} \, \Omega = 0.3 + 0.263 \, \Omega = 0.563 \, \Omega$$

The division by 4 is to allow for the ring nature of the circuit.

$$I_a = \frac{U_o}{Z_s} = \frac{240}{0.563} \, A = 426 \, A$$

We must then use the time/current characteristic of {Fig 3.13} to ascertain an operating time of 0.10 s.

From {Table 5.8} the value of k is 115.

$$\text{Then} \quad S = \frac{\sqrt{(I_a^2 t)}}{k} = \frac{\sqrt{(426^2 \times 0.10)}}{115} \, mm^2 = 1.17 \, mm^2$$

Since this value is smaller than the intended value of 1.5 mm^2, this latter value will be satisfactory.

Example 5.4

A 240 V single-phase circuit is to be wired in p.v.c. insulated single core cables enclosed in plastic conduit. The circuit length is 45 m and the live conductors are 16 mm^2 in cross-sectional area. The circuit will supply fixed equipment, and is to be protected by a 63 A HBC fuse to BS 88. The earth-fault loop impedance external to the installation has been ascertained to be 0.58 Ω. Calculate a suitable size for the circuit protective conductor.

With the information given this time the approach is somewhat different. We know that the maximum disconnection time for fixed equipment is 5 s, so from the time/current characteristic for the 63 A fuse {Fig 3.15} we can see that the fault current for disconnection will have a minimum value of 280 A.

$$\text{Thus,} \quad Z_s = \frac{U_O}{I_a} = \frac{240}{280} \ \Omega = 0.857 \ \Omega$$

If we deduct the external loop impedance, we come to the resistance of phase and protective conductors.

$$R_1 + R_2 = Z_s - Z_E = 0.857 - 0.58 \ \Omega = 0.277 \ \Omega$$

Converting this resistance to the combined value of R_1 and R_2 per metre,

$$(R_1 + R_2) \text{ per metre} = \frac{0.277 \times 1000}{45 \times 1.2} \ \text{m}\Omega/\text{m} = 5.13 \ \text{m}\Omega/\text{m}$$

Consulting {Table 5.5} we find that the resistance of 16 mm^2 copper conductor is 1.15 mΩ/m, whilst 10 mm^2 and 6 mm^2 are 1.83 and 3.08 mΩ/m respectively. Since 1.15 and 3.08 add to 4.23, which is less than 5.13, it would seem that a 6 mm^2 protective conductor will be large enough. However, to be sure we must check with the adiabatic equation.

$$R_1 + R_2 = \frac{(1.15 + 3.08) \times 1.2 \times 45}{1000} \ \Omega = 0.228 \ \Omega$$

$$Z_s = Z_E + (R_1 + R_2) = 0.58 + 0.228 \ \Omega = 0.808 \ \Omega$$

$$I_a = \frac{U_O}{Z_s} = \frac{240}{0.808} \ \text{A} = 297 \ \text{A}$$

From {Fig 3.15} the disconnection time for a 63 A fuse carrying 297 A is found to be 3.8 s.

From {Table 5.8} the value of k is 115.

$$\text{Then} \quad S = \frac{\sqrt{(I_a^2 t)}}{k} = \frac{\sqrt{(297^2 \times 3.8)}}{115} \ \text{mm}^2 = 5.03 \ \text{mm}^2$$

Since 5.03 is less than 6 then a 6 mm^2 protective conductor will be large enough to satisfy the requirements

5.4.6 *Unearthed metalwork* ~ [471-13]

If exposed conductive parts are isolated, or shrouded in non-conducting material, or are small so that the area of contact with a human body is limited, it is permissible not to earth them. Examples are overhead line metalwork which is out of reach, steel reinforcing rods within concrete lighting columns, cable clips, nameplates, fixing screws and so on. Where areas are accessible only to skilled or instructed persons, and where unauthorised persons are unlikely to enter due to the presence of warning notices, locks and so on, earthing may be replaced by the provision of obstacles which make direct contact unlikely, provided that the installation complies with the Electricity at Work Regulations, 1989.

5.5   Earth electrodes

5.5.1   *Why must we have earth electrodes?*
The principle of earthing is to consider the general mass of earth as a reference (zero) potential. Thus, everything connected directly to it will be at this zero potential, or above it by the amount of the volt drop in the connection system (for example, the volt drop in a protective conductor carrying fault current). The purpose of the earth electrode is to connect to the general mass of earth.

With the increasing use of underground supplies and of protective multiple earthing (PME) it is becoming more common for the consumer to be provided with an earth terminal rather than having to make contact with earth using an earth electrode.

5.5.2   *Earth electrode types*   ~   [514-13-01 and 542-02]
Acceptable electrodes are rods, pipes, mats, tapes, wires, plates and structural steelwork buried or driven into the ground. The pipes of other services such as gas and water must *not* be used as earth electrodes although they must be bonded to earth as described in {5.4.3}. The sheath and armour of a buried cable may be used with the approval of its owner and provided that arrangements can be made for the person responsible for the installation to be told if the cable is changed, for example, for a type without a metal sheath.

The effectiveness of an earth electrode in making good contact with the general masss of earth depends on factors such as soil type, moisture content, and so on. A permanently-wet situation may provide good contact with earth, but may also limit the life of the electrode since corrosion is likely to be greater. If the ground in which the electrode is placed freezes, there is likely to be an increase in earth resistance. In most parts of the UK an earth electrode resistance in the range $1\,\Omega$ to $5\,\Omega$ is considered to be acceptable.

The method of measuring the resistance of the earth electrode will be considered in {8.6.1}; the resistance to earth should be no greater than 220 Ω. The earthing conductor and its connection to the earth electrode must be protected from mechanical damage and from corrosion. Accidental disconnection must be avoided by fixing a permanent label as shown in {Fig 5.17} which reads

---

SAFETY ELECTRICAL CONNECTION
—
DO NOT REMOVE

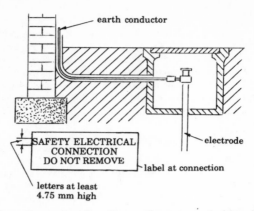

Fig 5.17   Connection of earthing conductor to earth electrode

5.6   Protective multiple earthing (PME)

5.6.1   *What is protective multiple earthing?*

If a continuous metallic earth conductor exists from the star point of the supply transformer to the earthing terminal of the installation, it will run throughout in parallel with the installation neutral, which will be at the same potential. It therefore seems logical that one of these conductors should be removed, with that remaining acting as a combined protective and neutral conductor (PEN). When this is done, we have a TN-C-S installation {5.2.4}. The combined earth and neutral system will apply only to the supply, and not to the installation.

Because of possible dangers with the system which will be explained in the following sub-sections, PME can be installed by the Electricity Supply Company only after the supply system and the installations it feeds have complied with certain requirements. These special needs will be outlined in {5.6.4}.

The great virtue of the PME system is that neutral is bonded to earth so that a phase to earth fault is automatically a phase to neutral fault. The earth-fault loop impedance will then be low, resulting in a high value of fault current which will operate the protective device quickly. It must be stressed that the neutral and earth conductors are kept quite separate within the installation: the main earthing terminal is bonded to the incoming combined earth and neutral conductor by the Electricity Supply Company. The difficulty of ensuring that bonding requirements are met on construction sites means that PME supplies must not be used. Electricity Supply Regulations forbid the use of PME supplies to feed caravans and caravan sites.

5.6.2   *Increased fire risk*

As with other systems of earth-fault protection, PME does not prevent a fault occurring, but will ensure that the fault protection device operates quickly when that fault appears. For example, if a fault of $2\,\Omega$ resistance occurs in a 240 V circuit protected by a 20 A semi-enclosed fuse in a system with an earth-fault loop impedance of $6\,\Omega$, the fault current will be 240/(2 + 6) A = 240/8 A = 30 A. The fuse would not blow unless the circuit were already loaded, when load current would add to fault current. If the circuit were fully loaded with a load current of 20 A, total current would be 50 A and the fuse would blow after about 18 s. During this time, the power produced in the fault would be:

$$P = I^2R = 30^2 \times 6 = 5400\,\text{W or }5.4\,\text{kW}$$

This could easily start a fire. If, however, the earth-fault loop impedance were 1 Ω, current would be 80 A and the fuse would blow in about 1.6 s and limit the energy in the fault circuit.

5.6.3 *Broken neutral conductor*

The neutral of a supply is often common to a large number of installations. In the (unlikely) event of a broken neutral, all the consumers on the load side of the break could have a combined neutral and earth potential of the same level as the phase system (240 V to earth). This situation could be very dangerous, because all earthed metalwork would be at 240 V above the potential of the general mass of earth.

To prevent such an event, the Electricity Supply Company connects its combined neutral and earth conductor to earth electrodes at frequent intervals along its run. Whilst this does not entirely remove the danger, it is much reduced. For example, the assumed earth resistance values of {Fig 5.18} show that the maximum possible potential to earth in this case would be 96 V. In practice, much lower resistance values for the earth connections will reduce this voltage. The Electricity Supply Company goes to very great lengths to ensure the integrity of its neutral conductor.

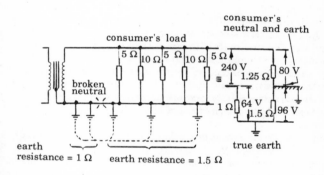

Fig 5.18   Danger due to broken neutral in a PME system

5.6.4 *Special requirements PME-fed installations* ~ [542-01-03 and 547-02-01]

An installation connected to a protective multiple earth supply is subject to special requirements concerning the size of earthing and bonding leads, which are generally larger in cross-section than those for installations fed by supplies with other types of earthing. Full discussions with the Electricity Supply Company are necessary before commencing such an installation to ensure that their needs will be satisfied. The cross-sectional area of the equipotential bonding conductor is related to that of the neutral conductor as shown in {Table 5.9}

Danger can arise when the non-current carrying metalwork of an installation is connected to the neutral, as is the case with a PME-fed system. The earth system is effectively in parallel with the neutral, and will thus share the normal neutral current. This current will not only be that drawn by the installation itself, but may also be part of the neutral current of neighbouring installations.

Table 5.9 Minimum cross-sectional area of main equipotential bonding conductor for PME-fed installations
(from [Table 54H] of BS 7671: 1992)

Neutral conductor c.s.a (mm²)	*Main equipotential bonding conductor c.s.a (mm²)*
35 or less	10
over 35 and up to 50	16
over 50 and up to 95	25

It follows that the earth system for an installation may carry significant current (of the order of tens of amperes) even when the main supply to that installation is switched off. This could clearly cause a hazard if a potentially explosive part of an installation, such as a petrol storage tank, were the effective earth electrode for part of the neutral current of a number of installations. For this reason, the Health and Safety Executive has banned the use of PME in supplies for petrol filling stations. Such installations must be fed from TN-S supply systems (HSE booklet HS(G)41 — 'Petrol Filling Stations: Construction and Operation').

5.7   Earthed concentric wiring
5.7.1   *What is earthed concentric wiring?*
This is the TN-C system {5.2.5} where a combined neutral and earth (PEN) conductor is used throughout the installation as well as for the supply. The PEN conductor is the sheath of a cable and therefore is concentric with (totally surrounds) the phase conductor(s). The system is unusual, but where employed almost invariably uses mineral insulated cable, the metallic copper sheath being the combined neutral and earth conductor.

5.7.2   *Requirements for earthed concentric wiring* ~ [546-01 and 546-02]
Earthed concentric wiring may only be used under very special conditions, which usually involve the use of a private transformer supply or a private generating plant. Since there is no separate path for earth currents, it follows that residual current devices (RCDs) will not be effective and therefore must not be used. The cross-sectional area of the sheath (neutral and earth conductor) of a cable used in such a system must never be less than 4 mm² copper, or 16 mm² aluminium or less than the inner core for a single core cable. All multicore copper mineral insulated cables comply with this requirement, even a 1 mm² two core cable having the necesssary sheath cross-sectional area. However, only single core cables of 6 mm² and below may be used. The combined protective and neutral conductors (sheaths) of such cables must not serve more than one final circuit.

Wherever a joint becomes necessary in the PEN conductor, the contact through the normal sealing pot and gland is insufficient; an extra earth tail must be used as shown in {Fig 5.19}. If it becomes necessary to separate the neutral and protective conductors at any point in an installation, they must not be connected together again beyond that point.

5.8   Other protection methods
Most of this chapter has been concerned with the protection of the user of the electrical installation against severe shock. The approach has not been to prevent shock altogether, because this is impossible, but to limit its se-

verity by ensuring that the shock current level is low and that the duration of its flow is very short. This section deals with the prevention of shock altogether by the use of special methods.

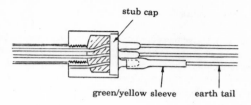

stub cap

green/yellow sleeve   earth tail

Fig 5.19   *Earth tail seal for use in earthed concentric wiring*

5.8.1   *Class II equipment* ~ [413-03 and 471-09]

Class II equipment has reinforced or double insulation. As well as the basic insulation for live parts, there is a second layer of insulation, either to prevent contact with exposed conductive parts or to make sure that there can never be any contact between such parts and live parts. The outer case of the equipment need not be made of insulating material; if protected by double insulation, a metal case will not present any danger. It must never be connected with earth, so connecting leads are two-core, having no protective conductor. The symbol for a double-insulated appliance is shown in {Fig 5.20}.

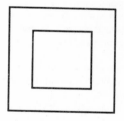

Fig 5.20   *British Standard symbol for double insulation*

To make sure that the double insulation is not impaired, it must not be pierced by conducting parts such as metal screws. Nor must insulating screws be used, because there is the possibility that they will be lost and will be replaced by metal screws. Any holes in the enclosure of a double-insulated appliance, such as those to allow ventilation, must be so small that fingers cannot reach live parts (IP2X protection). Class II equipment must be installed and fixed so that the double insulation will never be impaired, and so that metalwork of the equipment does not come into contact with the protective system of the main installation. Where the whole of an installation is comprised of Class II equipment, so that there is no protective system installed, the situation must be under proper supervision to make sure that no changes are made which will introduce earthed parts.

5.8.2   *Non-conducting location*   ~   [413-04, 471-10 and 514-13-02]
The non-conducting location is a special arrangement where there is no earthing or protective system because:-

1. there is nothing which needs to be earthed

2. exposed conductive parts are arranged so that it is impossible to touch two of them, or an exposed conducting part and an extraneous conductive part, at the same time. The distance between the parts must be at least 2 m, or 1.25 m if they are out of arm's reach. An alternative is to erect suitable obstacles, or to insulate the extraneous conductive parts.

Examples of extraneous conductive parts are water and gas pipes, structural steelwork, and even floors and walls which are not covered with insulating material. Insulation tests on floors and walls are considered in {8.5.2}. There must be no socket outlets with earthing contacts in a non-conducting location. This type of installation could cause danger if earthed metal were introduced in the form of a portable appliance fed by a lead from outside the location.

The potential reached by exposed metalwork within the situation is of no importance because it is never possible to touch two pieces of metalwork with differing voltage levels at the same time. Care must be taken, however, to make sure that a possible high potential cannot be transmitted ouside the situation by the subsequent installation of a conductor such as a water or gas pipe. A notice must be erected to state that a non-conducting location exists, and giving details of the person in charge who alone will authorise any work to be undertaken in, or will authorise any equipment to be taken into, the location. If two faults to exposed conductive parts occur from conductors at different potentials (such as a phase and a neutral) and there is a defective bonding system, dangerous potential differences could occur between exposed conductive parts. To prevent this possibility, double pole fuses or circuit breakers must provide overload protection in non-conducting locations.

Non-conducting locations are unusual, and their use must be limited to situations where there is continuous and proper supervision to ensure that the requirements are fully met and are properly maintained. This type of installation should only be considered after consulting a fully qualified electrical engineer.

5.8.3   *Earth-free bonding*   ~   [413-05, 471-11 and 514-13-02]
The Regulations permit the provision of an area in which all exposed metal parts are connected together, but not to earth. Inside the area, there can be no danger, even if the voltage to earth is very high, because all metalwork which can be touched will be at the same potential. Care is necessary, however, to prevent danger to people entering or leaving the area, because then they may be in contact with parts which are inside and others which are ouside the area, and hence at differing potentials. A notice must be erected to warn that the bonding conductors in the system must not be connected to earth, and that earthed equipment must not be brought into the situation (*see* {Table 5.10}).

As in the case of non-conducting locations, this type of installation is unusual, and must only be undertaken when designed and specified by a fully qualified electrical engineer. The area inside the protected system is often referred to as a 'Faraday cage'.

5.8.4 *Electrical separation* ~ [413-06, 471-12 and 514-13-02]

Safety from shock can sometimes be ensured by separating a system completely from others so that there is no complete circuit through which shock current could flow. It follows that the circuit must be small to ensure that earth impedances are very high and do not offer a path for shock current (*see* {Fig 5.3(b)}). The source of supply for such a circuit could be a battery or a generating set, but is far more likely to be an isolating transformer with a secondary winding providing no more than 500 V. Such a transformer must comply with BS 3535, having a screen between its windings and a secondary winding which has no connection to earth.

There must be no connection to earth, so a notice as shown in {Fig 5.10} must be posted and precautions must be taken to ensure, as far as possible, that earth faults will not occur. Such precautions would include the use of flexible cords without metallic sheaths, using double insulation, making sure that flexible cords are visible throughout their length of run, and so on. Perhaps the most common example of a separated circuit is the bathroom transformer unit feeding an electric shaver. By breaking the link to the earthed supply system using the double wound transformer, there is no path to earth for shock current (*see* {Fig 5.21}).

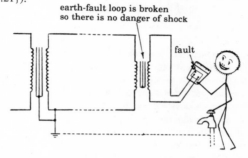

earth-fault loop is broken
so there is no danger of shock

fault

Fig 5.21   Bathroom shaver socket

5.9   Residual current devices (RCDs)

5.9.1   *Why do we need residual current devices?*

{5.3} has stressed that the standard method of protection is to make sure that an earth fault results in a fault current high enough to operate the protective device quickly so that fatal shock is prevented. However, there are cases where the impedance of the earth-fault loop, or the impedance of the fault itself, are too high to enable enough fault current to flow. In such a case, either:

1. current will continue to flow to earth, perhaps generating enough heat to start a fire, or
2. metalwork which is open to touch may be at a high potential relative to earth, resulting in severe shock danger.

Either or both of these possibilites can be removed by the installation of a residual current device (RCD).

In recent years there has been an enormous increase in the use of initials for residual current devices of all kinds. The following list, which is not exhaustive, may be helpful to readers:

RCD   residual current device
RCCD   residual current operated circuit breaker
SRCD   socket outlet incorporating an RCD
PRCD   portable RCD, usually an RCD incorporated into a plug
RCBO   an RCCD which includes overcurrent protection
SRCBO   a socket outlet incorporating an RCBO

5.9.2 *The principle of the residual current device* ~ [514-12-02]

The RCD is a circuit breaker which continuously compares the current in the phase with that in the neutral. The difference between the two (the residual current) *will* be flowing to earth, because it has left the supply through the phase and has not returned in the neutral (see {Fig 5.22}). There will always be some residual current in the insulation resistance and capacitance to earth, but in a healthy circuit such current will be low, seldom exceeding 2 mA.

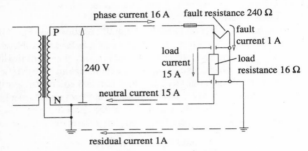

Fig 5.22   The meaning of the term residual current

The purpose of the residual current device is to monitor the residual current and to switch off the circuit quickly if it rises to a preset level. The arrangement of an RCD is shown in simplified form in {Fig 5.23}. The main contacts are closed against the pressure of a spring, which provides the energy to open them when the device trips. Phase and neutral currents pass through identical coils wound in opposing directions on a magnetic circuit, so that each coil will provide equal but opposing numbers of ampere turns when there is no residual current. The opposing ampere turns will cancel, and no magnetic flux will be set up in the magnetic circuit.

Residual earth current passes to the circuit through the phase coil but returns through the earth path, thus avoiding the neutral coil, which will therefore carry less current. This means that phase ampere turns exceed neutral ampere turns and an alternating magnetic flux results in the core. This flux links with the search coil, which is also wound on the magnetic circuit, inducing an e.m.f. into it. The value of this e.m.f. depends on the residual current, so it will drive a current to the tripping system which depends on the difference between phase and neutral currents. When the amount of residual current, and hence of tripping current, reaches a pre-determined level, the circuit breaker trips, opening the main contacts and interrupting the circuit.

For circuit breakers operating at low residual current values, an amplifier may be used in the trip circuit. Since the sum of the currents in the phases and neutral of a three-phase supply is always balanced, the system can be used just as effectively with three-phase supplies. In high current circuits, it is more usual for the phase and neutral conductors to simply pass through the magnetic core instead of round coils wound on it.

Operation depends on a mechanical system, which could possibly become stiff when old or dirty. Thus, regular testing is needed, and the RCD is provided with a test button which provides the rated level of residual current to ensure that the circuit breaker will operate. All RCDs are required to display a notice which draws attention to the need for frequent testing which can be carried out by the user, who presses a test button, usually marked T. {Table 5.10} shows the required notice.

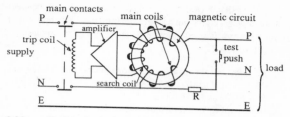

Fig 5.23   Residual current circuit breaker

The test circuit is shown in {Fig 5.23}, and provides extra current in the phase coil when the test button is pressed. This extra current is determined by the value of the resistor R.

Some RCDs (usually electronic types) will not switch off unless the mains supply is available to provide power for their operation. In such a case, mains failure may prevent tripping whilst danger is still present (due to, for example, charged capacitors). Such RCDs may only be used where there is another means of protection from indirect contact, or where the only people using the installation are skilled or instructed so that they are aware of the risk.

There are currently four basic types of RCD. Class AC devices are used where the residual current is sinusoidal — this is the normal type which is in the most wide use. Class A types are used where the residual current is sinusoidal and/or includes pulsating direct currents — this type is applied in special situations where electronic equipment is used. Class B is for specialist operation on pure direct current or on impulse direct or alternating current. Class S RCDs have a built-in time delay to provide discrimination (*see* below).

It must be understood that the residual current is the difference between phase and neutral currents, and that the current breaking ability of the main contacts is not related to the residual operating current value. There is a widely held misunderstanding of this point, many people thinking that the residual current setting is the current breaking capability of the device. It is very likely that a device with a breaking capacity of 100 A may have a residual operating current of only 30 mA.

There are cases where more than one residual current device is used in an installation; for example, a complete installation may be protected by an RCD rated at 100 mA whilst a socket intended for equipment outdoors may be protected by a 30 mA device. Discrimination of the two devices then becomes important. For example, if an earth fault giving an earth current of 250 mA develops on the equipment fed by the outdoor socket, both RCDs will carry this fault current, and both will become unbalanced. Since the fault is higher than the operating current of both devices, both will have their trip systems activated. It does not follow that the device with the smaller operating current will open first, so it is quite likely that the 100 mA device will operate, cutting off the supply to the complete installation even though the fault was on a small part of it. This is a lack of discrimination between the residual current devices. To ensure proper discrimination, the device with the larger operating current has a deliberate delay built into its operation. It is called a time delayed RCD.

5.9.3   *Regulations for residual current devices* ~ [412-06, 413-02-16 and 413-02-17, 471-08-06 and 471-08-07, 471-16, 514-12-02 and 531-02 to 531-04]

The primary purpose of the residual current device is to limit the severity of shock due to indirect contact. In other words, it will detect and clear earth faults which otherwise would could lead to dangerous potential differences between pieces of metalwork which are open to touch. If the sensitivity of the device (its operating residual current) is low enough, it may also be used to limit the shock received from direct contact in the case of the failure of other measures. A problem which may occur here is nuisance tripping, because the operating current may be so low that normal leakage current will cause operation. For example an RCD with a sensitivity of 2 mA will switch off the supply as soon as a shock current of 2 mA flows, virtually preventing a fatal shock. The difficulty is that normal insulation resistance leakage and stray capacitance currents can easily reach this value in a perfectly healthy system, and it may thus be impossible to keep the circuit breaker closed. The sum of the leakage currents in circuits protected by an RCD should never be more than 25% of the operating current of the device. Normal earth leakage current from equipments and appliances will, of course vary with the condition of the device. Maximum permitted leakage currents are listed in Appendix L of the 2nd Edition of Guidance Note 1, and vary from 0.25 mA for Class II appliances to 3.5 mA for information technology equipment (*see* {7.8.2})

Some RCDs (usually electronic types) will not switch off unless the mains supply is available to provide power for their operation. In such a case, mains failure may prevent tripping whilst danger is still present, (due to, for example, charged capacitors). Such RCDs may only be used where there is another means of protection from indirect contact, or where the only people using the installation are skilled or instructed so that they are aware of the risk.

If a residual current circuit breaker is set at a very low sensitivity, it can prevent death from electric shock entirely. However, the problem is that a safe current cannot be determined, because it will vary from person to person, and also with the time for which it is applied. The Regulations require a sensitivity of 30 mA for RCDs intended to provide additional protection from direct contact.

An RCD must not be used in an installation with neutral and earth combined (TN-C system using a PEN conductor) because there will be no residual current in the event of a fault to cause the device to operate, since there is no separate path for earth fault currents.

RCD protection is required for socket outlets where:

1.   they are part of a TT system (no earth terminal provided by the Electricity Supply Company),
2.   they are installed in a bedroom which contains a shower cubicle, or
3.   the socket outlet(s) are likely to feed portable equipment used outdoors.

Some RCDs are made so that the operating residual current may altered Such devices must not be installed where the setting could be altered by unauthorised persons.

Although residual current devices are current-operated, there are circumstances where the combination of operating current and high earth-fault loop impedance could result in the earthed metalwork rising to a dangerously high potential. The Regulations draw attention to the fact that if the product of operating current (A) and earth-fault loop impedance (Ω) excceds 50, the potential of the earthed metalwork will be more than 50 V above earth potential and hence dangerous. This situation must not be allowed to arise *(see* {Fig 5.24}).

RCDs must be tested to ensure correct operation within the required operating times. Such tests will be considered in {8.6.3}.

Special requirements apply to RCDs used to protect equipment having normally high earth leakage currents, such as data processing and other computer-based devices. These installations are considered in {7.8.2}.

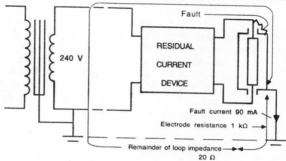

Fig 5.24   *Danger with an RCD when earth-fault loop impedance is high.*
In this case, p.d. from earth to exposed conductive parts will be
1000 Ω x 0.09 A = 90 V

5.9.4   Fault voltage operated circuit breakers

These circuit breakers, commonly known in the trade as 'voltage ELCBs', were deleted from the 15th Edition of the Wiring Regulations in 1985, and should not be installed. It is advisable that installations where they are still in use should be carefully tested prior to a change to residual current device protection.

5.10   Combined functional and protective earthing

The previous section has made it clear that high earth leakage currents can cause difficulties in protection. The increasing use of data processing equipment such as computers has led to the need for filters to protect against transients in the installation which could otherwise result in the loss of valuable data. Such filters usually include capacitors connected between live conductors and earth. This has led to large increases in normal earth currents in such installations, and to the need for special regulations for them. Since these are special situations, they will be considered in detail in {7.8}.

Electrical disturbances on the earth system (known as 'earth noise') may cause malfunctions of computer based systems, and 'clean' mains supplies and earth systems may be necessary. A separate earthing system may be useful in such a case provided that:
1.   the computer system has all accessible conductive parts earthed,
2.   the main earthing terminal of the computer earth system is connected directly to the main earthing terminal,
3.   all extraneous conductive parts within reach of the computer system are earthed to the main earthing terminal and not to the separate computer earth.

Supplementary bonding between the computer earth system and extraneous conductive parts is not necessary.

Circuits

6.1 Basic requirements for circuits [314-01]

The Regulations require that installations should be divided into circuits, the purposes being:

1. to prevent danger in the event of a fault by ensuring that the fault current is no greater than necessary to operate the protective system. For example, a large three-phase motor must be connected to a single circuit because the load cannot be subdivided. If, however, a load consisted of three hundred lamps, each rated at 100 W, it would be foolish to consider putting all this load onto a single circuit. In the event of a fault, the whole of the lighting would be lost, and the fault current needed to operate the protective device (single-phase circuit current would be 125 A at 240 V) would be high enough to cause a fire danger at the outlet where the fault occurred. The correct approach would be to divide the load into smaller circuits, each feeding, perhaps, ten lamps.

2. to enable part of an installation to be switched off for maintenance or for testing without affecting the rest of the system.

3. to prevent a fault on one circuit from resulting in the loss of the complete installation (*see* {3.8.6} on the subject of discrimination).

The number of final circuits will depend on the types of load supplied, and must be designed to comply with the requirements for overcurrent protection, switching and the current-carrying capacity of conductors. Every circuit must be separate from others and must be connected to its own overcurrent protective fuse or circuit breaker in a switch fuse, distribution board, consumer's unit, *etc. See* {Figs 6.1 and 6.2}.

A durable notice giving details of all the circuits fed is required to be posted in or near each distribution board. The data required is the equipment served by each circuit, its rating, its design current and its breaking capacity. When the occupancy of the premises changes, the new occupier must be provided with full details of the installation (*see* reference to the Operating Manual in {8.8.1}). These data must always be kept up to date.

6.2 Maximum demand and diversity

6.2.1 *Maximum demand* ~ [311-01]

Maximum demand (often referred to as MD) is the largest current normally carried by circuits, switches and protective devices; it does not include the levels of current flowing under overload or short circuit conditions. Assessment of maximum demand is sometimes straightforward. For example, the maximum demand of a 240 V single-phase 8 kW shower heater can be calculated by dividing the power (8 kW) by the voltage (240 V) to give a current of 33.3 A. This calculation assumes a power factor of unity, which is a reasonable assumption for such a purely resistive load.

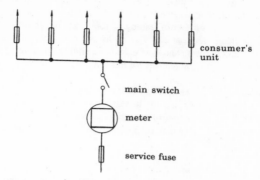

Fig 6.1   Typical arrangement for feeding final circuits in a domestic installation

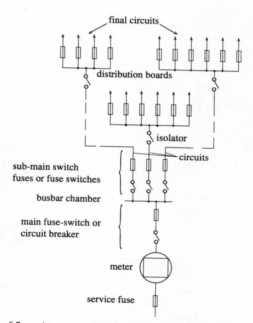

Fig 6.2   An arrangement for main and final circuits in a large installation

There are times, however, when assessment of maximum demand is less obvious. For example, if a ring circuit feeds fifteen 13 A sockets, the maximum demand clearly should not be 15 x 13 = 195 A, if only because the circuit protection will not be rated at more than 32 A. Some 13 A sockets may feed table lamps with 60 W lamps fitted, whilst others may feed 3 kW washing machines; others again may not be loaded at all. Guidance is given in {Table 6.1}.

Lighting circuits pose a special problem when determining MD. Each lampholder must be assumed to carry the current required by the connected load, subject to a minimum loading of 100 W per lampholder (a demand of 0.42 A per lampholder at 240 V). Discharge lamps are particularly difficult to assess, and current cannot be calculated simply by dividing lamp power by supply voltage. The reasons for this are:

1   control gear losses result in additional current,
2   the power factor is usually less than unity so current is greater, and
3   chokes and other control gear usually distort the waveform of the current so that it contains harmonics which are additional to the fundamental supply current.

So long as the power factor of a discharge lighting circuit is not less than 0.85, the current demand for the circuit can be calculated from:
$$\text{current (A)} = \frac{\text{lamp power (W)} \times 1.8}{\text{supply voltage (V)}}$$
For example, the steady state current demand of a 240 V circuit supplying ten 65 W fluorescent lamps would be:
$$I = \frac{10 \times 65 \times 1.8}{240} \text{ A} = 4.88 \text{ A}$$
Switches for circuits feeding discharge lamps must be rated at twice the current they are required to carry, unless they have been specially constructed to withstand the severe arcing resulting from the switching of such inductive and capacitive loads.

Table 6.1 Current demand of outlets

Type of outlet	Assumed current demand
2 A socket outlet	At least 0.5 A
Other socket outlets	Rated current
Lighting point	Connected load, with minimum of 100 W
Shaver outlet, bell transformer or any equipment of 5 W or less	May be neglected
Household cooker	10 A + 30% of remainder + 5 A for socket in cooker unit

When assessing maximum demand, account must be taken of the possible growth in demand during the life of the installation. Apart from indicating that maximum demand must be assessed, the Regulations themselves give little help. Suggestions for the assumed current demand of various types of outlet are shown in {Table 6.1}.

6.2.2   *Diversity*   ~   [311-01 and 433]
A domestic ring circuit typically feeds a large number of 13 A sockets but is usually protected by a fuse or circuit breaker rated at 30 A or 32 A. This means that if sockets were feeding 13 A loads, more than two of them in use at the same time would overload the circuit and it would be disconnected by its protective device.

In practice, the chances of all domestic ring sockets feeding loads taking 13 A is small. Whilst there may be a 3 kW washing machine in the kitchen, a 3 kW heater in the living room and another in the bedroom, the chance of all three being in use at the same time is remote. If they are all connected at the same time, this could be seen as a failure of the designer when assessing the installation requirements; the installation should have two ring circuits to feed the parts of the house in question.

Most sockets, then, will feed smaller loads such as table lamps, vacuum cleaner, television or audio machines and so on. The chances of all the sockets being used simultaneously is remote in the extreme provided that the number of sockets (and ring circuits) installed is large enough. The condition that only a few sockets will be in use at the same time, and that the loads they feed will be small is called *diversity*.

By making allowance for reasonable diversity, the number of circuits and their rating can be reduced, with a consequent financial saving, but without reducing the effectiveness of the installation. However, if diversity is over-estimated, the normal current demands will exceed the ratings of the protective devices, which will disconnect the circuits — not a welcome prospect for the user of the installation! Overheating may also result from overloading which exceeds the rating of the protective device, but does not reach its operating current in a reasonably short time. The Regulations require that circuit design should prevent the occurrence of small overloads of long duration.

The sensible application of diversity to the design of an installation calls for experience and a detailed knowledge of the intended use of the installation. Future possible increase in load should also be taken into account. Diversity relies on a number of factors which can only be properly assessed in the light of detailed knowledge of the type of installation, the industrial process concerned where this applies, and the habits and practices of the users. Perhaps a glimpse into a crystal ball to foresee the future could also be useful!

6.2.3   *Applied diversity*
Apart from indicating that diversity and maximum demand must be assessed, the Regulations themselves give little help. Suggestions of values for the allowances for diversity are given in {Table 6.2}.

Distribution boards must *not* have diversity applied so that they can carry the total load connected to them.

Example 6.1
A shop has the following single-phase loads, which are balanced as evenly as possible across the 415 V three-phase supply.

    2 x 6 kW and 7 x 3kW thermostatically controlled water heaters
    2 x 3 kW instantaneous heaters
    2 x 6 kW and 1 x 4 kW cookers
    12 kW of discharge lighting (sum of tube ratings)
    8 x 30 A ring circuits feeding 13 A sockets.

Calculate the total demand of the system, assuming that diversity can be applied.

Calculations will be based on {Table 6.2}.

Table 6.2 Allowance for diversity

Note the following abbreviations:
X is the full load current of the largest appliance or circuit
Y is the full load current of the second largest appliance or circuit
Z is the full load current of the remaining appliances or circuits

Type of final circuit	Types of premises		
	Households	*Small shops, stores, offices*	*Hotels, guest houses*
Lighting	66% total demand	90% total demand	75% total demand
Heating and power	100% up to 10 A + 50% balance	100%X + 75% (Y + Z)	100%X + 80%Y + 60%Z
Cookers	10 A + 30% balance + 5A for socket	100%X + 80%Y + 60%Z	100%X + 80%Y + 60%Z
Motors (but not lifts)		100%X + 80%Y + 60%Z	100%X + 50%(Y+Z)
Instantaneous water heaters	100%X + 100%Y + 25%Z	100%X + 100%Y + 25%Z	100%X + 100%Y +25%Z
Thermostatic water heaters	100%	100%	100%
Floor warming installations	100%	100%	100%
Thermal storage heating	100%	100%	100%
Standard circuits	100%X + 40%(Y+Z)	100%X + 50%(Y+Z)	100%X +50%(Y+Z)
Sockets and stationary equip.	100% X + 40 %(Y+ Z)	100 % X + 75% (Y+Z)	100 % X + 75 %Y+ 40%Z

The single-phase voltage for a 415V three-phase system is $415/\sqrt{3} = 240$ V.
All loads with the exception of the discharge lighting can be assumed to be at unity power factor, so current may be calculated from

$$I = \frac{P}{U}$$

Thus the current per kilowatt will be $\frac{1000}{240}$ A $= 4.17$ A

Water heaters (thermostatic)
No diversity is allowable, so the total load will be:
$$(2 \times 6) + (7 \times 3) \text{ kW} = 12 + 21 \text{ kW} = 33 \text{ kW}$$
This gives a total single-phase current of I $= 33 \times 4.17 = $ *137.6 A*

Water heaters (instantaneous)
100% of largest plus 100% of next means that in effect
there is no allowable diversity.
Single-phase current thus $= 2 \times 3 \times 4.17 = $ *25.0 A*

Cookers
100% of largest $=$ 6×4.17 A $= 25.0$ A
80% of second $= \frac{80}{100} \times 6 \times 4.17$ A $= 20.0$ A
60% of remainder $= \frac{60}{100} \times 4 \times 4.17$ A $= 10.0$ A

Total for cookers =   *55.0 A*

Discharge lighting
90% of total which must be increased to allow for
power factor and control gear losses.
Lighting current = $\frac{12 \times 4.17 \times 1.8 \times 90}{100}$ A   =   *81.1A*

Ring circuits
First circuit 100%, so current is   30 A
75% of remainder = $\frac{7 \times 30 \times 75}{100}$ =   157.5 A
Total current demand for ring circuits =   187.5A   *187.5A*

Total single phase current demand =   *486.2A*

Since a perfect balance is assumed, three phase line current = $\frac{486.2}{3}$ A

                                                     =   *162 A*

6.3   BS 1363 socket outlet circuits
The BS 1363 socket is the well known fused 13 A rectangular pin type
which has become the standard for domestic and commercial use in the UK
(*see* {Fig 6.3}).

6.3.1   The fused plug ~ [533-01 and 553-01]
In many situations there is a need for socket outlets to be closely spaced so
that they are available to feed appliances and equipment without the need to
use long and potentially dangerous leads. For example, the domestic kitchen
worktop should be provided with ample sockets to feed the many appliances (deep fat fryer, kettle, sandwich toaster, carving knife, toaster, micro-
wave oven, coffee maker, and so on) which are likely to be used. Similarly,
in the living room we need to supply television sets, video recorders, stereo
players, table lamps, room heaters, *etc*. In this case, more outlets will be
needed to allow for occasional rearrangement of furniture, which may well
obstruct access to some outlets.

    If each one of these socket outlets were wired back to the mains posi-
tion or to a local distribution board, large numbers of circuits and cables
would be necessary, with consequent high cost. The alternative is the provision of fewer sockets with the penalties of longer leads and possibly the use
of multi-outlet adaptors. Because the ideal situation will have closely-spaced
outlets, there is virtually no chance of more than a small proportion of them
being in use at the same time, so generous allowance can be made for diversity. Thus, cables and protective devices can safely be smaller in size than
would be needed if it were assumed that all outlets were simultaneously
fully loaded.

    Thus a ring circuit protected by a 30 A or 32 A device may well feed
twenty socket outlets. It follows that judgement must be used to make as
certain as possible that the total loading will not exceed the protective device rating, or its failure and inconvenience will result. Two basic steps will
normally ensure that a ring circuit is not overloaded.
1.   Do not feed heavy and steady loads (the domestic immersion heater
is the most obvious example) from the ring circuit, but make special provision for them on separate circuits.
2.   Make sure that the ring circuit does not feed too great an area. This
is usually ensured by limiting a single ring circuit to sockets within an area
not greater than one hundred square metres.

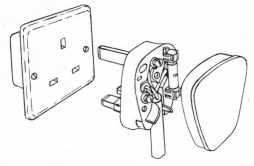

Fig 6.3   Plug and socket to BS 1363

We have already indicated that a 30 A or 32 A fuse or circuit breaker is likely to protect a large number of outlets. If this were the only method of protection, there could be a dangerous situation if, for example, a flexible cord with a rating of, say, 5 A developed a fault between cores. {Figure 3.13} shows that a 30 A semi-enclosed fuse will take 5 s to operate when carrying a current of almost 90 A, so the damage to the cord would be extreme. Because of this a further fuse is introduced to protect the appliance and its cord. The fuse is inside the BS 1363 plug, and is rated at 13 A or 3 A, although many other ratings up to 12 A, which are not recognised in the BSS, are available.

A plug to BS 1363 without a fuse is not available. The circuit protection in the distribution board or consumer's unit covers the circuit wiring, whilst the fuse in the plug protects the appliance and its cord as shown in {Fig 6.4}. In this way, each appliance can be protected by a suitable fuse, for example, a 3 A fuse for a table lamp or a 13 A fuse for a 3 kW fan heater.

Whilst the installer of the wiring is seldom concerned with the flexible cords of appliances connected to it, he must still offer guidance to users. This will include fitting 3 A fuses in plugs feeding low rated appliances, and the use of flexible cords which are of sufficient cross-section and are as short as possible in the cirumstances concerned. Generally, 0.5 mm^2 cords should be the smallest size connected to plugs fed by 30 A or 32 A ring circuits. Where the cord length must be 10 m or greater, the minimum size should be 0.75 mm^2 and rubber-insulated cords are preferred to those that are PVC insulated.

This type of outlet is not intended for use at high ambient temperatures. A common complaint is the overheating of a fused plug and socket mounted in an airing cupboard to feed an immersion heater; as mentioned above, it is not good practice to connect such a load to a ring circuit, and if unavoidable, final connection should be through a fused spur outlet.

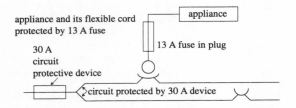

Fig 6.4   Principle of appliance protection by plug fuse

The British fused plug system is probably the biggest stumbling block to the introduction of a common plug for the whole of Europe (the 'europlug'). The proposed plug is a reversible two-pin type, so would not comply with the Regulations in terms of correct polarity. If we were to adopt it, every plug would need adjacent fuse protection, or would need to be rewired back to its own protective device. In either case, the cost would be very high.

Ring circuits fed from systems where no earth terminal is provided by the Electricity Supply Company (TT systems) must be protected by an RCD rated at 30 mA. In all installations, a socket intended to feed equipment outdoors must be individually protected by a 30 mA RCD.

Where a socket is mounted on a vertical wall, its height above the floor level or the working surface level level must be such that mechanical damage is unlikely. A minimum mounting height of 150 mm is reccommended.

6.3.2   The ring final circuit
The arrangement of a typical ring circuit is shown in {Fig 6.5} and must comply with the following requirements.
1 .   The floor area served by each ring must not exceed 100 m^2 for domestic situations. Where ring circuits are used elsewhere (such as in commerce or industry) the diversity must be assessed to ensure that maximum demand will not exceed the rating of the protective device.
2.   Consideration should be given to the provision of a separate ring (or radial) circuit in a kitchen.
3.   Where there is more than one ring circuit in the same building, the installed sockets should be shared approximately evenly between them.
4.   Cable sizes for standard circuits are as follows:
a)   p.v.c. insulated: 2.5 mm^2 for live (phase and neutral) conductors and 1.5 mm^2 for the CPC
b)   mineral insulated: 1.5 mm^2 for live conductors and 1 mm^2 for the CPC
       These sizes assume that sheathed cables are clipped direct, are embedded in plaster, or have one side in contact with thermally insulating material. Single core cables are assumed to be enclosed in conduit or trunking. No allowance has been made for circuits which are bunched, and the ambient temperature is assumed not to exceed 30°C.
5.   The number of unfused spurs fed from the ring circuit must not exceed the number of sockets or fixed appliances connected directly in the ring.
6.   Each non-fused spur may feed no more than one single or one twin socket, or no more than one fixed appliance.
7.   Fixed loads fed by the ring must be locally protected by a fuse of rating no greater than 13 A or by a circuit breaker of maximum rating 16 A.
8.   Fixed equipment such as space heaters, water heaters of capacity greater than 15 litres, and immersion heaters, should not be fed by a ring, but provided with their own circuits.

6.3.3   The radial circuit
Two types of radial circuit are permitted for socket outlets. In neither case is the number of sockets to be supplied specified, so the number will be subject to the constraints of load and diversity. The two standard circuits are:

1.   20 A fuse or miniature circuit breaker protection with   2.5 mm^2 live and 1.5 mm^2 protective conductors (or 1.5 mm^2 and 1.0 mm^2 if m.i. cable) feeding a floor area of not more than 20 m^2. If the circuit feeds a kitchen or utility room, it must be remembered that a 3 kW device such as a washing machine or a tumble dryer takes 12.5 A at 240 V and that this leaves little

capacity for the rest of the sockets.

2.   32 A cartridge fuse to BS88 or miniature circuit breaker feeding
through 4 mm² live and 2.5 mm² protective conductors (or 2.5 mm² and 1.5
mm² if m.i. cable) to supply a floor area no greater than 50 m².

The arrangement of the circuits is shown in {Fig 6.6}. 6 mm² may seem
to be a large cable size in a circuit feeding 13 A sockets. It must be remembered, however, that the 2.5 mm² ring circuit allows current to be fed both
ways round the ring, so that two conductors are effectively in parallel, whereas
the 4 mm² cable in a radial circuit must carry all the current.

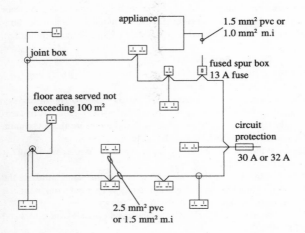

Fig 6.5   Ring circuit feeding socket outlets to BS 1363

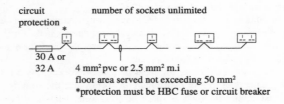

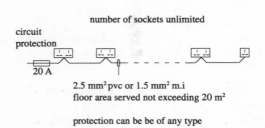

Fig 6.6   Radial circuits

120

Radial circuits can be especially economic in a long building where the completion of a ring to the far end could effectively double the length of cable used. As for ring circuits, danger can occur if flexible cords are too small in cross-section, or are too long, or if 3 A fuses are not used where appropriate.

The minimum cross-sectional area for flexible cords should be:
0.5 mm^2 where the radial circuit is protected by a 16 A fuse,
0.75 mm^2 for a 20 A fuse,
or 1.0 mm^2 for a 30 A or 32 A fuse.

6.4 Industrial socket outlet circuits

6.4.1 Introduction

There is no reason at all to prevent the installation of BS 1363 (13 A) socket outlets in industrial situations. Indeed, where light industry, such as electronics manufacture, is concerned, these sockets are most suitable. However, heavy duty industrial socket outlets are available, and this type is the subject of this section.

6.4.2 BS 196 socket outlet circuits ~ [537-05, 553-01 and 553-02]

BS 196 sockets are two-pin, non-reversible, with a scraping earth connection. The fusing in the plug can apply to either pole, or the plug may be unfused altogether. Interchangeability is prevented by means of a keyway which may have any of eighteen different positions, identified by capital letters of the alphabet (*see* {Fig 6.7}). They are available with current ratings of 5 A, 15 A or 30 A.

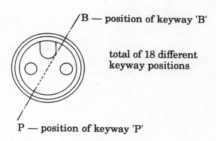

B — position of keyway 'B'

total of 18 different keyway positions

P — position of keyway 'P'

Fig 6.7   *Arrangement of socket outlet to BS 196. Two of the eighteen possible keyway positions are shown*

Circuit details for wiring BS 196 outlets can be summarised as:

1. the maximum protective device rating is 32 A
2. all spurs must be protected by a fuse or circuit breaker of rating no larger than 16 A -this means that 30 A outlets cannot be fed from spurs
3. the number of sockets connected to each circuit is unspecified, but proper judgement must be applied to prevent failure of the protective device due to overload
4. cable rating must be no less than that of the protective device for radial circuits, or two thirds of the protective device rating for a ring circuit
5. on normal supplies with an earthed neutral, the phase pole must be fused and the keyway must be positioned at point B (*see* {Fig 6.7})

6. when the socket is fed at reduced voltage from a transformer with the centre tap on its secondary winding earthed {Fig 6.8}, both poles of the plug must be fused and the keyway must be positioned at P.

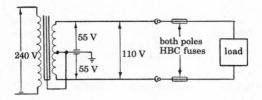

Fig 6.8   *Circuit fed from a transformer with a centre-tapped secondary winding*

6.4.3   ***BS EN 60309-2 (BS 4343) socket outlet circuits*** ~ [537-05, 553-01 and 553-02]

Plugs and sockets to BS EN 60309-2 are for industrial applications and are rated at 16 A, 32 A, 63 A and 125 A. All but the smallest size must be wired on a separate circuit, but 16 A outlets may be wired in unlimited numbers on radial circuits where diversity can be justified. However, since the maximum rating for the protective device is 20 A, the number of sockets will be small except where loads are very light or where it is certain that few loads will be connected simultaneously. An arrangement of BS EN 60309-2 plugs and sockets is shown in {Fig 6.9}.

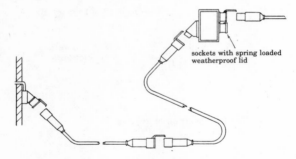

sockets with spring loaded weatherproof lid

Fig 6.9   *Plugs and sockets to BS EN 60309-2*

6.5   **Other circuits**
6.5.1   ***Lighting circuits*** ~ [422-01, 553-03 and 553-04, 554-01 and 554-02]

Lampholders and ceiling roses must not be used in installations where the supply voltage exceeds 250 V. Where bayonet cap (BC) or edison screw (ES) lampholders are used, the protective device rating is limited to the values shown in {Table 6.3}, unless the lampholders and the associated wiring are enclosed within a fireproof enclosure, such as a luminaire (lighting fitting), or unless they have separate overcurrent protection in the form of a local fuse or circuit breaker.

Lampholders are often mounted within enclosed spaces such as lighting fittings, where the internal temperature may become very high, particularly where filament lamps are used. Care must be taken to ensure that the lampholders, and their associated wiring, are able to withstand the temperature concerned. Where ES lampholders are connected to a system with the neutral at earth potential (TT or TN systems) care must be taken to ensure that the centre contact is connected to the phase conductor and the outer screw to the neutral to reduce the shock danger in the event of touching the outer screw during lamp changing (*see* {Fig 6.10}).

Table 6.3 Overcurrent protection of lampholders
(from [Table 55B] of the 16th edition of BS 7671: 1992)

Type of lampholder		Maximum rating of protective device (A)
Bayonet cap	SBC	6
	BC	16
Edison screw	SES	6
	ES	16
	GES	16

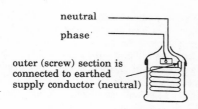

neutral

phase

outer (screw) section is
connected to earthed
supply conductor (neutral)

Fig 6.10   Correct connection of ES lampholder

Ceiling roses must not have more than one flexible cord connected to them, and, like the flexible cords themselves, must not be subjected to greater suspended weight than their design permits (*see* {Table 4.2}). Lampholders in bath or shower rooms must be fitted with a protective shield to prevent contact with the cap whilst changing the lamp (*see* {Fig 6.11}).

In large lighting installations, particularly where fluorescent fittings are involved, consideration should be given to the use of luminaire support couplers (LSCs) or plugs and sockets. Such arrangements facilitate the disconnection of luminaires for electrical maintenance and for cleaning, and may also allow the complete testing of an installation before erection of the luminaires. Many lighting installations are now controlled by sophisticated software (which may switch off the lighting when daylight levels increase or when a room has been unoccupied for a predetermined time). Such devices must be installed to comply with the Regulations.

6.5.2   *Cooker circuits*   ~   [476-03-04]

A cooker is regarded as a piece of fixed equipment unless it is a small table-mounted type fed from a plug by a flexible cord. Such equipment must be under the control of a local switch, usually in the form of a cooker control unit. This switch may control two cookers, provided both are within 2 m of it. In many cases this control unit incorporates a socket outlet, although often such a socket is not in the safest position for use to supply portable appliances, whose flexible cords may be burned by the hotplates. It is often considered safer to control the cooker with a switch and to provide a separate socket circuit. The protective device is often the most highly rated in an installation, particularly in a domestic situation, so there is a need to ensure that diversity has been properly calculated (*see* {6.2.2}).

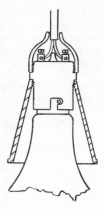

Fig 6.11   Protective shield for a BC lampholder

The diversity applicable to the current demand for a cooker is shown in {Table 6.2} as 10 A plus 30% of the remainder of the total connected load, plus 5 A if the control unit includes a socket outlet. A little thought will show that whilst this calculation will give satisfactory results under most circumstances, there is a danger of triggering the protective device under some circumstances. For example, at Christmas it is quite likely that both ovens, all four hotplates and a 3 kW kettle could be simultaneously connected. Just imagine the chaos which a blown fuse would cause! This alone is a very good reason for being generous with cable and protective ratings.

Example 6.2
A 240 V domestic cooker has the following connected loads:

top oven	1.5 kW
main oven	2.5 kW
grill	2.0 kW
four hotplates	2.0 kW each

The cooker control unit includes a 13 A socket outlet. Calculate a suitable rating for the protective device.

The total cooker load is $1.5 + 2.5 + 2.0 + (4 \times 2.0)$ kW = 14 kW

Total current = $\dfrac{P}{U}$ = $\dfrac{14000}{240}$ A = 58.3 A

The demand is is made up of:

the first 10 A   10.0 A

+ 30% of remainder = $\dfrac{30 \times (58.3 - 10)}{100}$ = $\dfrac{30 \times 48.3}{100}$ =   14.5 A

+ allowance for socket outlet   <u>5.0 A</u>

total   =   29.5 A

A 30 A protective device is likely to be chosen. The cable rating will depend on correction factors (*see* {Chapter 5}).

6.5.3   Off-peak appliance circuits
All Electricity Supply Companies offer extremely economic rates for energy taken at off-peak times, usually for seven hours each night (economy 7). The supply meter is usually arranged so that off-peak inexpensive energy can only be obtained from a special pair of terminals, whilst a second pair

provide energy throughout the 24 hour period, charging for it in terms of the times at which it is taken. In most cases, energy used at the cheap rate must be stored for use at other times. There are two major methods of storing energy, n both cases involving its conversion to heat.

1.   *storage heaters,* which are used for space heating. Circuits feeding them should always be wired radially, with only one flexible outlet for each. This will help to avoid problems in the event of the storage heater being changed or one of a different power rating.

2.   *immersion heaters,* the energy being stored as hot water in a lagged tank for use during the day. Since the amount of hot water used is variable, it is usually necessary to have a method of increasing the water temperature should that heated at the cheap rate be used up. This involves the use of a second immersion heater, or a single heater with a double element. Since convection, and hence water heating, takes place mainly above the active heater, an immersion heater placed low in the tank and fed by the off-peak supply will heat the whole tank, whilst a second heater, placed higher in the tank and connected to the normal supply, will be switched on when necessary to top up the temperature of the hot water stored {Fig 6.12(a)}. Sometimes a top mounted dual heater is used for the same purpose as shown in {Fig 6.12(b)}. The normal and off-peak heaters must be supplied through totally separate circuits. 3 kW heaters must be connected permanently to a double pole switch and not fed via a plug and socket.

6.6   Circuit segregation

6.6.1   *Circuit definitions and purposes* ~ [Part 2, 331-01 & 512-05]

Some types of circuit need to be kept separate from others:

1.   to ensure that extra-low voltage systems cannot have low voltage impressed on them,

2.   so that services which must be separated from others to make sure that they can operate correctly in an emergency (safety services such as fire detection and alarm circuits, and emergency lighting) will remain unaffected for as long as possible when fire occurs, and

3.   to prevent electromagnetic interference from the mains cables to telecommunication and other systems (electro-magnetic compatibility or EMC).

Circuits can be divided into four categories for these purposes.

Category 1
Any low voltage circuit which is not part of a fire alarm or emergency lighting system and is fed from the mains supply.

Category 2
Any extra-low voltage circuit other than those for fire alarms and emergency lighting. This would include telephones, radio sets, intruder alarms, sound systems, data transmission, bell and call systems, *etc.*

Category 3
Fire alarm circuits.

Category 4
Emergency lighting circuits.

The Regulations do not recognise Category 4 circuits, listing fire alarm and emergency lighting circuits as Category 3. This seems illogical, since the two types of circuit included in the Regulations as Category 3 must be segregated from each other.

A separation of at least 50 mm should be maintained between telecommunications and main cables, unless the latter are metal sheathed or are enclosed in conduit or in trunking or separated from telecommunications cables by a non-conducting divider. A particular problem may arise where the electrical installation shares a space occupied by a hearing aid induction loop system. In this situation, there care must be taken to keep phases and neutral, together with switch feeds and switch wires, as close together as possible to avoid setting up a mains voltage loop.

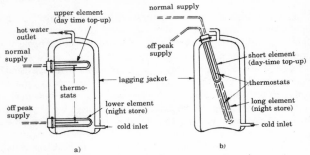

Fig 6.12   Arrangement of immersion heaters for off-peak supplies

6.6.2   Segregating circuits ~ [528-01]

The Regulations allow four possible methods of enclosing cables so that segregation is acceptable, and they are shown in {Fig 6.13}.

The methods of segregation are

1. partitioned trunking, with all four categories of circuit separated from each other. The partitions separating Category 3 and 4 cables from oth ers must be fire resistant
2. totally separate conduits, ducts or trunking for each of the four types of circuit
3. all four types of circuit sharing the same enclosure, provided that:
   a) Category 2 circuits are insulated to Category 1 levels, and
   b) Category 3 and 4 circuits are wired in mineral insulated cable
4. separate enclosures for Category 3 cables, Category 4 cables, and Cat egories 1 and 2, which can be together provided that Category 2 cables are insulated for the same voltage level as Category 1.

Where cables enter common outlet boxes they must be segregated by parti-tions to continue the requirements given; metal partitions must be earthed. Multicore cables may be used, but

1. Category 3 or Category 4 cores must never share a cable with those of the other three Categories
2. Where cores of Categories 1 and 2 share a common flexible cable, the Category 2 cores must be insulated for the same voltage as those of Category 1.

The IEE has published a Table showing the required separation of power and signal cables to comply with EMC requirements. As yet the separations are not widely used as they are considered too large to be practicable.

Table 6.4 Proposed EMI cable separation distances			
Power cable voltage	Min. separation btw power & signal cables, m	Power cable current	Min separation btw power & signal cables, m
115 V	0.25	5 A	0.24
240 V	0.45	15 A	0.35
415 V	0.58	50 A	0.50
3.3 kV	1.10	100 A	0.60
6.6 kV	1.25	300 A	0.85
11 kV	1.40	600 A	1.05

6.6.3   *Lift and hoist shaft circuits*   ~   [528-02-06]

A lift shaft may well seem to be an attractive choice for running cables, but this is not permitted except for circuits which are part of the lift or hoist installation (*see*{Fig 6.14}). The cables of the lift installation may be power cables fixed in the shaft to feed the motor(s), or control cables which feed call buttons, position indicators, and so on. Trailing cables will feed call buttons, position indicators, lighting and telephones in the lift itself.

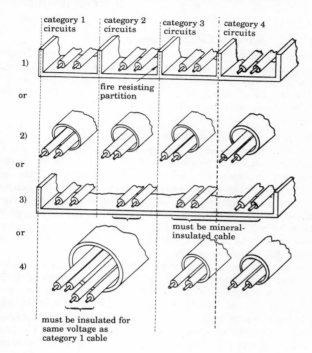

Fig 6.13   *Segregation of circuits*

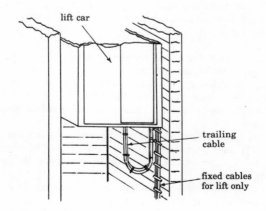

Fig 6.14   *Cables installed in lift or hoist shafts*

Special installations

7.1   Introduction [600-01 and 600-02]

The Regulations apply to all electrical installations in buildings. There are some situations out-of-doors, as well as special indoor ones, which are the subject of special requirements due to the extra dangers they pose, and these will be considered in this chapter. These Regulations are additional to all of the other requirements, and not alternatives to them.

7.2   Bath tubs and shower basins

7.2.1   *Introduction*   ~ [601-01 and 601-10]

People using a bathroom are often unclothed and wet. The absence of clothing (particularly shoes) will remove much of their protection from shock (*see*{3.4}), whilst the water on their skin will tend to short circuit its natural protection. Thus, such people are very vulnerable to electric shock due to their reduced body resistance, so special measures are needed to ensure that the possibility of contact (either direct or indirect) is much reduced. Attention is drawn to the fact that a bath used for medical treatment may need special consideration, because the possible hazard is greater.

Appliances and sockets for use in these high-risk areas must be separatedextra-low voltage (SELV) type, at a potential not exceeding 12 V; the exception is a shaver unit fed from a double wound transformer to BS 3052 (*see* {5.8.4}). Such equipment must be protected to level IP2X, which means that it must be impossible to touch live parts with a finger, and must have insulation capable of withstanding a voltage level of 500 V r.m.s. for one minute. Such socket outlets, or fused connection units in a bathroom or a bedroom containing a shower cubicle, must be protected by an RCD with an operating current no greater than 30 mA.

The special requirements of this section do not apply to rooms (such as bedrooms) containing an enclosed and prefabricated shower basin, provided that switches are not mounted within 0.6 m or sockets within 2.5 m of the door opening as indicated in {Fig 7.1}.

7.2.2   *Bath and shower room requirements*   ~ [601-02 to 601-09, 601-11 and 601-12]

Usually, protection against direct contact will be by means of earthed equipotential bonding and automatic disconnection of the supply in the event of a fault (the same as for most other installations), and the special requirements are:

1. all extraneous conductive parts must be bonded together and to earth (*see* {Fig 7.2}), and with no exposed metallic conductors. The exception to this is any SELV system, which must remain unconnected to the main earthing system

2. the earth fault loop impedance must be low enough to allow disconnection within 0.4 s (*see* {5.3}).

In no case within a bathroom is it permissible to rely for protection against direct contact on obstacles, placing out of reach, a non-conducting location or earth-free equipotential bonding. Switches must be out of reach of a person using the bath or shower, although cord-pull and similar remotely controlled switches may be used.

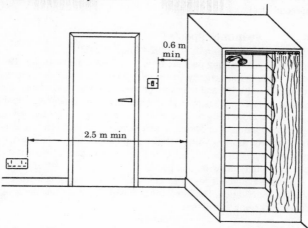

Fig 7.1   Permissible positions of switches and socket outlets in a room containing a shower cubicle but no bath

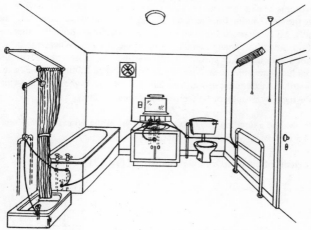

Fig 7.2   Bonding requirements for a bathroom [601-04-02]— the line showing bonding connections does not represent the actual positions of the cable

There must be no switchgear, control gear or accessories installed within the bath or shower basin, whilst wiring must not be metallic sheathed, or enclosed in metallic conduit or trunking if run on the surface. The installation of equipment within a bath may sound ridiculous, but jaccuzi pumps and devices to assist disabled people to get into and out of a bath are not uncommon. Where, as is often the case, controls are mounted on the bath, the whole system must be fed at no more than 12 V.

Lampholders must be provided with a protective shield (*see* {Fig 6.11}) if within 2.5 m of the bath, or totally-enclosed luminaires must be used. Electrical equipment installed beneath a bath (an example is a jacuzzi pump) must only be accessible after the use of tools. If electric floor heating is used in a room containing a bath or a shower cubicle, it must have its metal sheath, or a covering metal grid if there is no metallic sheath, connected to the local equipotential bonding.

7.3 Swimming pools

7.3.1 *Introduction* ~ [602-01 and 602-02]

People using a swimming pool are often partly unclothed and wet. The absence of clothing (particularly shoes) will remove much of their protection from shock (*see*{3.4}), and the water on their skin will tend to short circuit its natural protection. Thus, such people are very vulnerable to electric shock due to their reduced body resistance, so special measures are needed to ensure that the possibility of contact (either direct or indirect) is much reduced. Only separatedextra-low voltage (SELV) equipment may be installed, other than water heaters in zones B and C, and equipment specially designed to be safe in the vicinity of swimming pools.

Special Regulations for electrical installations in swimming pools are new in the 16th Edition. The Regulations classify zones around the pool, the arrangements being as follows:

Zone A

is the inside of the pool, including chutes and flumes, as well as apertures in the pool walls and floor which are accessible to the bathers

Zone B

is a volume above the pool to 2.5 m above the rim, plus the same height above the surrounding floor area on which people may walk, extending horizontally 2.0 m outwards from the rim. If the pool rim is above the surrounding floor level, the zone extends 2.5 m above the rim. Where the pool is provided with diving or spring boards, starting blocks or a chute, the zone also includes the volume enclosed by a vertical plane 1.5 m from the edge of the board and extending upwards to 2.5 m above the highest surface which will be used by people, or by the ceiling if there is one.

Zone C

is the volume extending 1.5 m horizontally from the boundary of Zone B and 2.5 m vertically above the floor.

The extents of the three zones are shown in {Fig 7.3}.

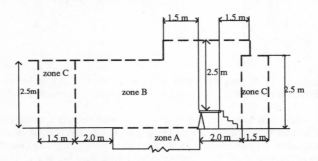

Fig 7.3   Dimensions of zones for swimming pools

7.3.2 *Special requirements for swimming pools* ~ [602-03 to 602-08]

Appliances and sockets for use in these high-risk areas must be separated extra-low voltage (SELV) type, at a potential not exceeding 12 V ac or 30 V dc. Such equipment must be protected to level IP2X, which means that it must be impossible to touch live parts with a finger, or must have insulation capable of withstanding a voltage level of 500 V r.m.s. for one minute. The safety source (a transformer in most cases), must be installed outside zones A, B and C. An exception is the supply to floodlights (usually they will be under water) which can be fed at up to 18 V, each floodlight being powered by a separate transformer or a separate secondary winding of a transformer having multiple secondary windings.

Usually, protection against direct contact will be by means of earthed equipotential bonding and automatic disconnection of the supply in the event of a fault (the same as for most other installations), and the special require-ments are:

1. all extraneous conductive parts, including non-insulating floors, may be bonded together and to earth. Solid floors in Zones B and C must have an equipotential bonded grid buried within them. The exception to this is any SELV system, which must remain unconnected to the main earthing system
2. the earth fault loop impedance must be low enough so as to allow discon-nection within 0.4 s (*see* {5.3}).

In no case is it permissible to rely for protection against direct contact on obstacles, placing out of reach, a non-conducting location or earth-free equipotential bonding. If wiring is run on the surface it must not be metallic sheathed and must not be run in metal conduit or metal trunking.

Enclosures used for wiring systems or appliances at swimming pools are subject to special requirements, which depend on whether water jets will be used for cleaning purposes. If they *will* be used, protection must be to IPX8, (submersion in water), within the pool (Zone A) and IPX5, which means that there must be effective protection against such water jets (*see* {Table 2.1}). If water jets will *not* be used, protection depends on the zone (*see* {7.3.1}) in which the system or appliance is situated. The requirements are:

Zone A
IPX8, which means protection from submersion in water,
Zone B
IPX4, so that protection is provided against splashing water, and
Zone C
indoor pools, IPX2, which gives protection against dripping water when inclined at 15°; and outdoor pools, IPX4, which gives protection against splashing water.

There must be no switchgear, control gear or accessories installed within zones A or B, except for small pools where socket outlets are necessary and cannot possibly be installed outside zone B. In this case, BS 4343 socket outlets may be installed provided they are at least 1.25 m (arms' reach) outside zone A, are at least 0.3 m above the floor, and are protected by an RCD with a 30 mA rating or are protected by electrical separation [413-05] with the necessary isolating transformer situated outside zones A, B and C. Wiring must not be metallic sheathed, or enclosed in metallic conduit or trunking if run on the surface. Joint boxes may be used, but they must be inaccessible if installed in zones A or B. Instantaneous water heaters com-plying with BS 3456 may be installed in zones B and C.

Socket outlets (in zone C only — but see above for small pools) must be to industrial standard BS 4343, and a shaver outlet to BS 3052 may also be installed in zone C {5.8.4}. If electric floor heating is used around a pool, it must have its metal sheath, or a covering metal grid if there is no metallic sheath, connected to the local equipotential bonding. Audio and public address equipment must be at least 3.5 m from the pool edge and must not be allowed to get wet. Loudspeakers must be out of reach of anyone in the water or at the poolside, and if portable microphones are used they must be connected by way of a suitable isolating transformer. Radio microphones will reduce the danger of shock.

7.4 Sauna rooms

7.4.1 *Introduction* ~ [603-01 and 603-02]

A sauna is a room in which the air is heated to a high temperature, humidity usually being very low, but occasionally increasing when water is deliberately poured over the heater. People using a sauna are usually unclothed and often wet (mainly due to perspiration), the absence of clothing (particularly shoes) removing much of their protection from shock (*see*{3.4.2}).

Special Regulations for saunas are new in the 16th Edition. As with bathrooms and swimming pools, zones are again classified, but this time they are concerned more with temperature levels than contact with water.

Zone A

The volume within 0.5 m horizontally from the sauna heater, and extending from the floor up to within 0.3 m from the ceiling.

Zone B

The volume covering the whole of the sauna room outside zone A up to 0.5 m above the floor.

Zone C

The volume directly above zone B and extending upwards to within 0.3 m from the ceiling.

Zone D

The volume covering the whole floor area of the room and extending down from the ceiling for a distance of 0.3 m, including the space in this volume directly above the sauna heater.

The extents of the four zones are shown in {Fig 7.4).

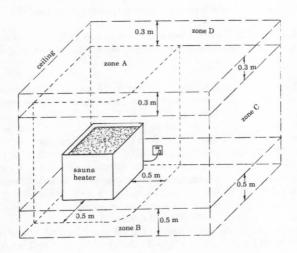

Fig 7.4   Definition of zones for sauna rooms

7.4.2   *Special requirements for saunas*   ~   [603-3 to 603-11]

If separated extra-low voltage (SELV) is used, protection against direct contact must be provided by:

1. enclosures or barriers providing protection to IP24. This means protec
tion against entry of human fingers and from splashing water, and
2. insulation which will withstand a voltage of 500 V r.m.s. for one minute.

     Some of the standard measures for protection against both direct and indirect contact must *not* be used in saunas. These are:

1. obstacles,
2. placing out of reach,
3. non-conducting location, and
4. earth free equipotential bonding.

     All equipment must be protected to at least IP24, and no equipment other than the sauna heater may be installed in zone A. There must be no socket outlets in a sauna room, nor other switchgear which is not built into the sauna heater. In the lower part of the room where it will not be so hot (zone B) there is no special requirement concerning the heat resistance of equipment. If installed in zone C, equipment must be suitable for operation at an ambient temperature of 125°C, and only luminaires, mounted so as to prevent them from overheating, are allowed in zone D. Any flexible cord used must be mechanically protected and insulated with 150°C rubber.

7.5   Installations on construction sites

7.5.1   *Introduction*   ~   [604-01 to 604-03]

The electrical installation on a construction site is there to provide lighting and power to enable the work to proceed. By the very nature of the situation, the installation will be subjected to the kind of ill treatment which is unlikely to be applied to most fixed installations. Those working on the site may be ankle deep in mud and thus particularly susceptible to a shock to earth, and they may be using portable tools such as drills and grinders in situations where danger is more likely than in most factory situations.

     Installations will also, by definition, be temporary. As the construction proceeds they will be moved and altered. It is usual for such installations to be subjected to thorough inspection and testing at intervals which will never exceed three months.

     As well as the erection of new buildings, the requirements for construction sites will also apply to:

1. sites where repairs, alterations or additions are carried out
2. demolition of buildings
3. public engineering works
4. civil engineering operations, such as road buiding,coastal protection, *etc.*

     The special requirements for construction sites do not apply to temporary buildings erected for the use of the construction workers, such as offices, toilets, cloakrooms, dormitories, canteens, meeting rooms, *etc.* These situations will not change as construction progresses, and are thus subject to the general requirements of the Regulations.

     The equipment used must be suitable for the particular supply to which it is connected, and for the duty it will meet on site. Where more than one voltage is in use, plugs and sockets must be non-interchangeable to prevent misconnection. Six levels of voltage are recognised for a construction site installation. They are:

1. 25 V single-phase SELV for portable hand-lamps in damp and confined situations,

2. 50 V single-phase, centre-point earthed for hand lamps in damp and confined situations,

3. 400 V three phase, for use with fixed or transportable equipment with a load of more than 3750 Watts,

4. 230 V single phase, for site buildings and fixed lighting,

5. 110 V three phase, for transportable equipment with a load up to 3750 Watts, and

6. 110 V single phase, fed from a transformer, often with an earthed centre-tapped secondary winding, to feed transportable tools and equipment, such as floodlighting, with a load of up to 2 kW. This supply ensures that the voltage to earth should never exceed 55 V (*see* {Fig 7.5}). The primary winding of the transformer must be RCD-protected unless the equipment fed is to be used indoors.

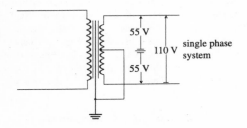

Fig 7.5   Arrangement of transformer for safety 110 V supply for a construction site

Supplies will normally be obtained from the Electricity Supply Company. Where a site is remote, so that a generator must be used (IT supply system) special protective requirements apply which are beyond the scope of this Guide, and the advice of a qualified electrical engineer must be sought.

7.5.2   *Special regulations for construction sites*   ~   [604-04 to 604-13]

Construction site installations are like most others in that they usually rely on earthed equipotential bonding and automatic disconnection for protecting from electric shock. This is the system where an earth fault, which results in metalwork open to touch becoming live, also causes a fault current which will open the protecive device to remove the supply within 0.4 s for socket outlet circuits or 5 s for fixed appliances. For construction sites, where the prospective dangers are greater, there are additional requirements. These are:

1. The times within which disconnection must occur are reduced, except for fixed equipment. Generally the times permitted are much reduced to reflect the more dangerous nature of the construction site. These voltage-related disconnection times are given in {Table 7.1} for TN systems, which are those where the Electricity Supply Company provides an earthing terminal. Where values of maximum earth-fault loop impedance are necessary to check compliance with this requirement, it must be appreciated that the values of {Tables 5.1 and 5.2} no longer apply, and new values must be calculated using the stated supply voltages, and disconnection currents read from {Figs 3.13 to 3.19} using the maximum times from {Table 7.1}. For 240 V circuits with a maximum disconnection time of 0.2 s, maximum permissible earth-

fault loop values for various types of protective device are shown in {Table 7.2}. Where the given values of earth-fault loop impedance cannot be met, protection must be by means of RCDs with operating current not exceeding 30 mA.

For fixed installations the disconnection time is 5 s, so {Tables 5.2 and 5.4} can be used to find maximum earth-fault loop impedance values.

2. In other installations the maximum shock voltage is given as 50 V, calculated from the impedance of the protective system to earth in ohms multiplied by the fault current in amperes. This was the basis of {Table 5.3}, but these data do not apply in this case. For construction sites, the value is reduced to 25 V. For example, for a TN system, the impedance of the circuit protective conductor (Z_s) multiplied by the current rating of the protective fuse or circuit breaker (I_n) must not exceed 25.

Table 7.1 Maximum disconnection times for construction site circuits (TN systems)

(from [Table 604A] of BS 7671: 1992)

Supply voltage (U_0) (volts)	Disconnection time (seconds)
120	0.35
220 to 277	0.20
400 and 480	0.05

For a circuit protected by a 15 A miniature circuit breaker type 1, which must carry a current of 60 A to trip in 5 s (*see* {Fig 3.16}), the impedance of the circuit protective conductor must therefore be no greater than:

$$Z_S = \frac{25}{I_n} = \frac{25}{60} \ \Omega = 0.42 \ \Omega$$

Sockets on a construction site must be separated extra-low voltage (SELV) or protected by a residual current circuit breaker (RCD) with an operating current of not more than 30 mA, or must be electrically separate from the rest of the supply, each socket being fed by its own individual transformer. SELV is unlikely for most applications, because 12 V power tools would draw too much current to be practical. Most sockets are likely to be fed at 110 V from centre-tapped transformers so will comply with this requirement.

Distribution and supply equipment must comply with BS 4363, and, together with the installation itself, must be protected to IP44. This means provision of mechanical protection from objects more than 1 mm thick and protection from splashing water. Such equipment will include switches and isolators to control circuits and to isolate the incoming supply. The main isolator must be capable of being locked or otherwise secured in the 'off' position.

Cables and their connections must not be subjected to strain, and cables must not be run across roads or walkways without mechanical protection. Circuits supplying equipment must be fed from a distribution assemby including overcurrent protection, a local RCD if necessary, and socket outlets where needed. Socket outlets must be enclosed in distribution assemblies, fixed to the outside of the assembly enclosure, or fixed to a vertical wall. Sockets must not be left unattached, as is often the case on construction sites. Socket outlets and cable couplers must be to BS EN 60309-2. A typical schematic diagram for a construction site system is shown in {Fig 7.6}.

Table 7.2 Maximum earth-fault loop impedance values for 240 V construction site circuits to give a maximum 0.2 s disconnection time
(from [Tables 604B1 and 604B2] of BS 7671: 1992)

Type of protection	Protection rating (A)	Max. loop impedance (Ω)
Cart. fuse, BS 1361	5	9.60
	15	3.00
	20	1.55
	30	1.00
Cart. fuse BS 88 pt 2	6	7.74
	10	4.71
	16	2.53
	20	1.60
	25	1.33
	32	0.92
MCB type 1	5	12.0
	10	6.00
	15	4.00
	20	3.00
	30	2.00
MCB type 2	5	6.86
	10	3.43
	15	2.29
	20	1.71
	30	1.14
MCB type 3	5	4.80
	10	2.40
	15	1.60
	20	1.20
	30	0.80
MCB type B	6	8.0
	10	4.80
	16	3.00
	20	2.40
	32	1.50

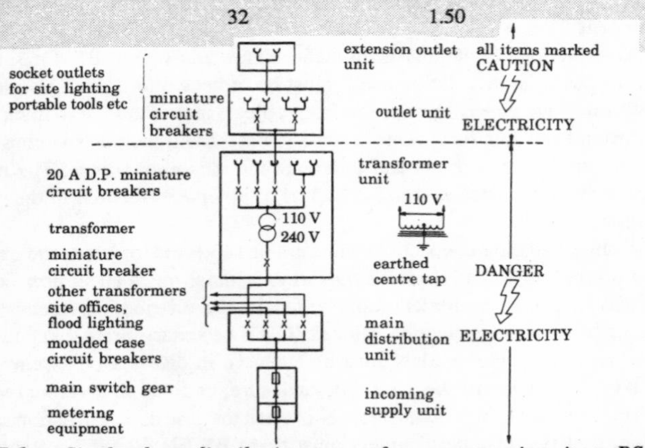

Fig 7.6   Single-phase distribution system for a construction site to BS 4363 and CP 1017

7.6 Agricultural and horticultural installations

7.6.1 *Introduction* ~ [605-01]

Situations of these kinds will pose an increased risk of electric shock because people and animals are more likely to be in better contact with earth than in other installations. Animals are particularly vulnerable, because their body resistance is much lower than for humans, and applied voltage levels which are quite safe for people may well prove fatal for them. If animals are present, the electrical installation may be subjected to mechanical damage (animals cannot be instructed to keep away from installations) as well as to a greater corrosion hazard due to the presence of animal effluents.

The special requirements for agricultural and horticultural installations also apply to locations where livestock are kept, such as stables, chicken houses, piggeries, etc. They also apply to feed processing locations, lofts, and storage places for hay, straw and fertilizers.

For these reasons, special Regulations apply to such installations, both indoors and outdoors. It is important to recognise that dwelling houses on agricultural and horticultural premises used solely for human habitation are excluded from these special requirements.

7.6.2 *Agricultural installations* ~ [605-02 to 605-13]

Of paramount importance where livestock are present is the indication that the levels of separated extra-low voltage (SELV) are too high. Whilst the Regulations do not suggest a safe voltage for animals, simply suggesting a level 'appropriate to the type of livestock', a practical value is likely to be no higher than 25 V. Such systems, as well as those complying with SELV requirements, must be protected to IP2X (human fingers must not be able to touch live parts) or the insulation must be able to withstand 500 V r.m.s. for one minute.

All socket outlets must be protected by residual current device(s) (RCDs) with an operating current of no more than 30 mA. Whilst it is accepted that livestock cannot be protected by earthed equipotential bonding and automatic disconnection (sometimes known as EEBAD) because the voltages to which they would be subjected in the event of a fault are unsafe for them, changes to standard installation requirements do offer some additional protection. The requirements are:

1. Disconnection times for the operation of protective devices are reduced, usually to half the normal value. The maximum times are voltage related, and are shown in {Table 7.3}.

Table 7.3 Maximum disconnection times for agricultural circuits for livestock (TN systems)
(from [Table 605A] of BS 7671: 1992)

Supply voltage (U_0) (volts)	Disconnection time (seconds)
120	0.35
220 to 277	0.20
400 and 480	0.05

The application of these reduced connection times leads to the reduced levels of maximum earth-fault loop impedance, shown in {Table 7.4} for 240 V circuits.

2. Fixed equipment and distribution circuits are permitted to have a disconnection time of 5 s, which is the same as for normal installations, and require the maximum earth-fault loop impedance values shown in {Tables 5.2 and

5.4}. An exception is where the fixed equipment is fed from the same distribution board as circuits requiring disconnection in 0.2 s at 240 V. In such a case

Either the resistance of the main protective conductor from the distribution board to the point of connection to the main equipotential bonding system must be low enough to ensure that its volt drop when fault current flows does not exceed 25 V,

or the distribution board must have its exposed conductive parts bonded to all extraneous conductive parts (such as water pipes) in the area.

3. The maximum of 25 V for the potential difference across the protective conductor under fault conditions stated above is applied to all final circuits. This is half of the level accepted in other installations, so the protective conductor resistance must have half the normal value. Note that where an IT system (usually a generating plant) is used, special requirements apply — these are outside the scope of this Guide, and advice must be sought from a qualified electrical engineer.

4. Supplementary equipotential bonding must be applied to connect together all exposed and extraneous conductive parts which are accessible to livestock and the main protective system. It is recommended that a metallic grid should be laid in the floor and connected to the protective conductor.

The equipotential bonding required will create an earth zone, and special measures are necessary where a circuit fed from this zone extends outside it. If the equipment could be touched by a person in contact with the general mass of earth, disconnection time must not exceed 0.2 s, even for fixed equipment. This means that {Table 7.4} will apply to such circuits.

Fire is a particular hazard in agricultural premises where there may be large quantities of loose straw or other flammable material. The Regulations require the protection of the system by an RCD with an operating current not greater than 500 mA. In practice, a 300 mA rating is likely to be used. This will result in problems of discrimination between this unit and those of lower operating current rating unless the main RCD is of the time delayed type (*see* {5.9.3}). Care must be taken to ensure that heaters are not in positions where they will ignite their surroundings; a clearance of at least 500 mm is required for radiant heaters.

All electrical equipment must be protected to IP44, and chosen to be suitable to operate under the onerous conditions they will experience. Wiring must be inaccessible to livestock and must be vermin proof. In practice, this will probably mean enclosure in galvanised steel conduit, or the use of mineral insulated cables. Switch and control gear must be to IP44 and constructed of, or enclosed in, insulating material. Switches for emergency use must not be in positions accessible to cattle, or where cattle may make operation difficult. The likelihood of panic amongst animals when emergencies occur must be taken into account.

7.6.3   *Electric fence controllers* ~ [605-14]

Electric fences are in wide use to prevent animals from straying. In most cases they are fed with very short duration pulses of a voltage up to 5 kV — however, the energy involved is too small to cause dangerous shock. Usually animals very quickly learn that it can be painful to touch a fence, and give it a wide berth. {Fig 7.7} indicates a typical arrangement for an electric fence, the numbers in brackets in the following text relating to the circled numbers on the diagram.

Controllers may be mains fed or battery operated. Controllers must be to BS EN 61011-1, and not more than one controller may be connected to a fence (1). So that it will give a short, sharp shock to animals, the high voltage output of the controller is connected between the fence and an earth electrode. It is of obvious importance that the high voltage pulses are not transmitted to the earthed system of a normal electrical installation because the earth zones overlap (4). A method of checking the earth zone of an electrode is described in {8.6.1}. For similar reasons it is important that the fence should never make contact with other wiring systems (2), telephone circuits, radio aerials, *etc*; the installer must also take account of induction from overhead lines, which may occur when the fence runs for a significant distance below such a line.

Table 7.4 Maximum earth-fault loop impedance values for 240 V agricultural circuits to give a maximum time of 0.2 s for disconnection
(from [Tables 605B1 and 605B2] of BS 7671: 1992)

Type of protection	Protection rating (A)	Max loop impedance (Ω)
Cart. fuse BS1361	5	9.60
	15	3.00
	20	1.55
	30	1.00
Cart. fuse BS88 pt 2	6	7.74
	10	4.71
	16	2.53
	20	1.60
	25	1.33
	32	0.92
MCB type 1	5	12.00
	10	6.00
	15	4.00
	20	3.00
	30	2.00
MCB type 2	5	6.86
	10	3.43
	15	2.29
	20	1.71
	30	1.14
MCB type 3	5	4.80
	10	2.40
	15	1.60
	20	1.20
	30	0.80

A mains operated fence controller must be installed so that interference by unauthorised persons and mechanical damage is unlikely. It must not be fixed to a pole carrying mains or telecommunication circuits, except that where it is fed by an overhead line consisting of insulated conductors, it may be fixed to the pole carrying the supply (3).

7.6.4 *Horticultural installations* ~ [605-02-02, 605-03, and 605-05 to 605-13]

A horticultural installation is likely to be subject to the same wet and high-earth-contact conditions experienced in agriculture, but will be free from the extra hazards associated with the presence of animals. In other respects it is subject to the same requirements as an agricultural installation as described in {7.6.2}. A type of load not so likely to be present in the agricultural situation is the soil warming circuit often used in horticulture. The relevant Regulations for such circuits are considered in {7.11.4}.

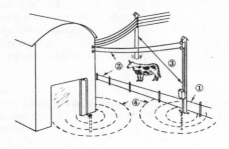

Fig 7.7   Electric fence controller

7.7   Restrictive conductive locations

7.7.1   *Introduction*   ~   [606-01]

A restrictive conductive location is one in which the surroundings consist mainly of metallic or conductive parts, with which a person within the location is likely to come into contact with a substantial portion of his body, and where it is very difficult to interrupt such contact. An example could be a large metal enclosure, such as a boiler, in which work must be carried out. Clearly, such a situation could result in considerable danger to a worker, who may be lying inside the boiler whilst using an electric drill or grinder.

7.7.2   *Special requirements*   ~   [606-02 to 606-04]

The person in the restrictive location is protected by the use of separatedextralow voltage (SELV) or functional extra-low voltage (FELV) systems. A SELV supply must be fed from a safety source, such as a special step-down isolating transformer. If the supply is FELV, the voltage level(s) must be

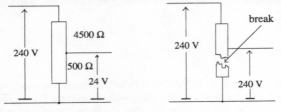

Fig 7.8   The danger when using a potential divider to provide extra-low voltage supplies

the same, but the supply need not be from a safety source, and could be obtained from an auto-transformer, a semiconductor, such as a thyristor or triac (*see* {Fig 3.5}), or a potential divider. In this case it is not impossible (due to failure in the source, such as an open circuit in part of the potential divider (*see*{Fig 7.8})) for mains voltage to appear on the conductors of the equipment in use. For this reason, protection against touching a live part (direct contact) must be provided by:

    either   enclosure providing protection to IP2X so that a human finger cannot touch live parts,

    or   insulation which will withstand 500 V r.m.s.for one minute.

Protection against direct contact by obstacles or placing out of reach is not permitted.

Protection against indirect contact (touching parts which have become live as the result of a fault) for fixed equipment is achieved by one of the following methods:

1. if the supply is SELV, then protection from indirect contact is automatically prevented because the system is totally isolated and has no earthing connections,

2. if the supply is FELV, the whole of the conductive location must be solidly earthed by the provision of a supplementary equipotential bonding conductor connecting it to the earthing system,

3. for both SELV and FELV protection can be by automatic disconnection, the protective device opening the circuit within 0.4 s in the event of a fault. To ensure low earth-fault loop impedance, the conductive location must again be solidly earthed,

4. when either SELV or FELV is used, protection by electrical isolation may be used. This means that equipment in use is fed from the unearthed secondary winding of an isolating transformer so that there is no path for earth currents. When more than one supply to equipment is required, each must be from a separate transformer secondary winding,

5. equipment to Class II must be used, with additional protection from an RCD with an operating current not exceeding 30 mA.

Separated and isolating sources, other than those specified in 3) above, must be situated outside the restrictive conductive location unless they are part of an installation in a permanent restrictive conductive location. By its very nature an enclosed space is likely to be dark, and handlamps will be needed by the operator. Socket outlets intended to feed handlamps must be SELV, and these outlets can also be used to feed hand tools. However, the limitation of 25 V on SELV means that the range of hand tools available is very limited, and where higher voltages are used, sockets for this purpose can be fed using electrical separation. Where equipment requires a functional earth connection, a socket may be fed using FELV.

7.8   Earthing for combined functional and protective purposes

7.8.1   *Introduction* ~ [545, 546, 607-01, 607-02-01 and 607-02-02]
Data processing equipment is extremely common. Such computer based equipments are used in the office for word processing, handling accounts, dealing with wages and so on. In the factory they control processes, and in the retail shop they are used for stock control, till management and many other purposes. Even in the home we are starting to see their use for security, for temperature control, for domestic banking services, *etc.* Telecommunications equipment is also becoming more widely used.

All such equipments have a common danger of failing and losing their stored data if subject to mains disturbances such as voltage spikes and transients. They are protected from such failures by feeding the supplies to them *via* filter circuits, which are designed to remove or reduce such voltage variations before they reach the sensitive circuitry. A simple filter is shown in {Fig 7.9} and will almost always include resistive and capacitive components which are connected from live conductors to earth. This will give rise to increased earth currents, driven by supply voltage through the resistive and reactive components. When the circuit, including the filter, is switched on, higher earth currents will usually flow for a very short time whilst capacitors are charging.

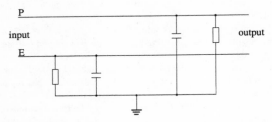

Fig 7.9   Mains transient suppression filter

Data processing and telecommunications equipment of this kind therefore has high levels of earth current in normal use, although in most cases the earth current produced by filters will not exceed 3.5 mA. Consequently, special regulations apply to the protective and earthing conductors of such circuits.

7.8.2   Special regulations for equipment with high earth currents
            ~   607-02-03 to 607-02-07 and 607-03 to 607-05]

Electricity Supply Regulation 26 indicates that the level of earth leakage current should not normally exceed one ten thousandth part of the installation maximum demand (for example, 10 mA earth leakage current for an installation with a maximum demand of 100 A). Data processing equipment is likely to have a higher leakage current than this, so special regulations become necessary. Foremost is the requirement that where earthing is used for functional purposes (to allow the filters to do their job) as well as protective purposes, the protective function must take precedence.

Stationary equipment with an earth leakage current exceeding 3.5 mA must be permanently connected, or an industrial plug and socket to BS 4343 must be used. When a socket outlet circuit may be expected to feed data processing equipment with normal earth leakage current of more than 10 mA, or if the earth leakage current for a circuit feeding fixed stationary equipment is greater than 10 mA, earthing must be through a high integrity protective system complying with at least one of the following:

1. a protective conductor of cross-sectional area at least 10 mm^2
2. duplicated protective conductors having separate connections and each of at least 4 mm^2 cross-sectional area
3. duplicated protective cross-sectional areas of all the conductors is at least 10 mm^2, in which case the metallic sheath, armour or braid of the cable may be one of the protective conductors, provided that it complies with the adiabatic equation of [543-01-03] (*see* {5.4.4})
4. duplicated protective conductors, one of which can be metal conduit, trunking or ducting, whilst the other is a 2.5 mm^2 conductor installed within it
5. an earth monitoring device is used which switches off the supply automatically if the protective conductor continuity fails (*see* [543-03-05])
6. connection of the equipment to the supply through a double wound transformer, a protective conductor complying with one of the arrangements 1 - 5 above connecting exposed conductive parts to a point on the secondary winding.

The reason for these precautions is that if the circuit protective conductor should become open circuit, leakage current could flow to earth through a person touching exposed conductive parts with possibly lethal consequences.

Alternatively to items 1 - 6 above, a ring circuit may be used provided that it feeds single sockets, has no spurs, and that the protective conductor, which must be no smaller than 1.5 mm², has its two ends separately connected at the earthing block of the fuse or circuit breaker board.

Where an installation is protected by an RCD the sum of the earth leakage currents due to data processing equipment must not exceed 25% of the device tripping current. Where this requirement cannot be met, connection must be *via* a double wound transformer as in item 6 above. For other installations, no specific figure for the leakage current as a percentage of tripping current is given; the requirement is that the normal leakage current will be unlikely to cause unnecessary tripping [531-6]. The 25% limit means effectively either that a single RCD with a high operating current must be used, or that the installation must be subdivided to allow a number of lower rated RCDs to be used. If the Electricity Supply Company does not provide an earthed terminal and an installation electrode is required (TT system), the result of multiplying the total earth leakage current in amperes and *twice* the resistance of the earth electrode in ohms must not exceed 50 (volts). For normal TT systems, there is no need to double the earth electrode resistance, so in the case of data processing equipment the earth electrode resistance must effectively be half its value for other installations. Data processing equipment must not be connected to a system which is not earthed in the normal way (IT system).

7.9 Caravan, motor caravan and caravan park installations

7.9.1 *Introduction* ~ [608-01]

A caravan is a leisure accommodation vehicle which reaches its site by being towed by a vehicle. A motor caravan is used for the same purpose, but has an engine which allows it to be driven; the accommodation module on a motor caravan may sometimes be removed from the chassis. Caravans will often contain a bath or a shower, and in these cases the special requirements for such installations (*see*{7.2}) will apply. Railway rolling stock is not included in the definition as a caravan. Caravans used as mobile workshops will be subject to the requirements of the Electricity at Work Regulations 1989 as well as the IEE Wiring Regulations.

All the dangers associated with fixed electrical installations are also present in and around caravans. Added to these are the problems of moving the caravan, including connection and disconnection to and from the supply, often by totally unskilled people. Earthing is of prime importance because the dangers of shock are greater. For example, the loss of the main protective conductor *and* a fault to the metalwork in the caravan is likely to go unnoticed until someone makes contact with the caravan whilst standing outside it {Fig 7.10}.

7.9.2 *Caravan installations* ~ [608-02 to 608-08]

The requirements for the electrical installation are listed below, numbers relating to those shown on {Fig 7.11}.

1. An inlet coupler to BS EN 60309-2 with its keyway at position 6h must be provided no higher than 1.8 m above the ground,

2. A spring hinged lid which will close to protect the coupler socket when travelling must be fitted,

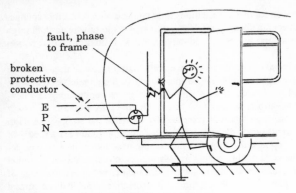

Fig 7.10   Importance of earthing a caravan

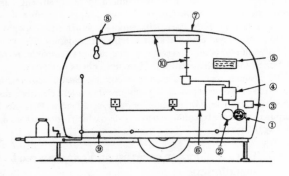

Fig 7.11   Requirements for the installation in a caravan

3. A clear and durable notice must be provided beside the inlet coupler to indicate the voltage, frequency and current of the caravan installation,

4. Protection by automatic disconnection (as for all other installations) but using double pole MCBs to disconnect all live conductors, together with a double pole RCD with an operating current of 30 mA and means of isolating the complete caravan installation must be used,

5. A durable notice must be fixed beside the isolator with the wording shown in {Table 7.5},

6. All circuits must be provided with a protective conductor. All sockets are to be three pin with earthed contact and must have no accessible conductive parts. If two or more systems at different voltages are in use, the plugs of the differing systems must not be interchangeable. ELV sources must be 12 V, 24 V or 48 V when dc, or 12 V, 24 V, 42 V or 48 V when ac. Such ELV sockets must have their voltage clearly marked.Wiring may be flexible or or with at least seven strands in non-metallic conduit (pliable polyethylene conduits must not be used) or sheathed flexible cables, with the smallest conductor being 1.5 mm^2 in cross-sectional area. Where 240 V and extra-low voltage circuits (usually 12 V) are both used, the cables of the two systems must be run separately. Since the wiring will be subjected to vibration when the caravan is moved, great care must be taken to ensure that bushes or grommets are used where it passes through metalwork,

7. Enclosed luminaires must be fixed directly to the structure or lining of the caravan. Where luminaires are of the dual voltage type (240 V mains and

12 V battery supply) they must be fitted with separate and different types of lampholder, with proper separation between wiring of the two supplies, and be clearly marked to indicate the lamp wattages and voltages. They must be designed so that both lamps can be illuminated at the same time without causing damage.

8. Pendant luminaires must have arrangements to secure them whilst the caravan is being moved.

9. All metal parts of the caravan, with the exception of metal sheets forming part of the structure, must be bonded together and to a circuit protective conductor, which must not be smaller than 4 mm^2 except where it forms part of a sheathed cable or is enclosed in conduit.

10. Sheathed cables must be supported at intervals of at least 400 mm where run vertically, or 250 mm horizontally unless run in non-metallic rigid conduit.

11. No electrical equipment may be installed in a compartment intended for the storage of gas cylinders.

Table 7.5 Instructions for electricity supply

To connect

1	Before connecting the caravan installation to the mains supply, check that: a) the supply available at the caravan pitch supply point is suitable for the caravan installation and appliances, and b) the caravan main switch is in the OFF position
2	Open the cover to the appliance inlet provided at the caravan supply point and insert the connector of the supply flexible cable
3	Raise the cover from the electricity outlet provided on the pitch supply point and insert the plug of the supply cable.

The caravan supply flexible cable must be fully uncoiled to avoid damage by overheating

4	Switch ON at the caravan main switch
5	Check the operation of residual current devices, if any, fitted in the caravan by depressing the test button.

In case of doubt or if, after carrying out the above procedure the supply does not become available, or if the supply fails, consult the caravan park operator or his agent or a qualified electrician

To disconnect

6	Unplug both ends of the cable.

Periodical inspection

Preferably not less than once every three years and more frequently if the vehicle is used more than the normal average mileage for such vehicles, the caravan electrical installation and supply cable should be inspected and tested and a report on its condition obtained as prescribed in BS 7671 (formerly the Regulations for Electrical Installations) published by the Institution of Electrical Engineers.

Where a caravan appliance may be exposed to the effects of moisture, it must be protected to IP55 (protected from dust and from water jets). Every caravan which includes an electrical installation must be provided with a flexible lead not more than 25 m long fitted with a BS EN 60309-2 plug and a BS EN 60309-2 connector with the keyway at position 6h. The cross-sectional area of the cable must be related to the rated current of the plug as shown by {Table 7.6}.

Table 7.6 Cross-sectional areas of flexible cables and cords for supplying caravan connectors
(from [Table 608A] of BS 7671: 1992)

Rated current of plug (A)	Cross-sectional area (mm²)
16	2.5
25	4.0
32	6.0
63	16.0

7.9.3 *Caravan park installations* ~ [608-09 to 608-13]

These Regulations cover the arrangements for the supply of electrical energy to individual plugs on a caravan or tent park. The plots on which caravans will stand are referred to as *caravan pitches*.

Wherever possible, the supplies to the plugs should be by means of underground cables (*see* {7.13.3}). These cables should be installed outside the area of the caravan pitch to ensure that they are not damaged by tent pegs (which are often used in erecting caravan awnings) and ground anchors which are used to secure the caravans against the effects of high winds. If the underground cables must be below the pitches, they must be provided with additional protection, as shown in {Fig 7.22}.

If overhead supplies are used they must:

1. be constructed with insulated, rather than bare, cables
2. always be at least 2 m outside the area of every pitch
3. have a mounting height of at least 3.5 m, which must be increased to 6 m where vehicle movements are possible. Since in most cases the whole of the area of a caravan site is subject to vehicle movements, most over head systems will need to be at a minimum height of 6 m. Poles or other supports for overhead wiring must be located or protected so that they are unlikely to be damaged by vehicle movement.

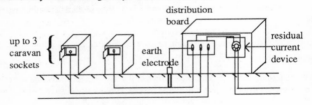

Fig 7.12   RCD protection for sockets at caravan pitches

There must be at least one supply socket for each pitch, the socket positioned so that it is between 0.8 m and 1.5 m above ground level, and no more than 20 m from any point in its pitch. The current rating must be at least 16 A, and the outlet must be to BS EN 60309-2, of the splash proof type to IPX4, and with the keyway at position 6h. If the expected current demand for a pitch exceeds 16 A, extra sockets of higher rating must be installed. A problem frequently arises due to the increasing use of 3 kW instantaneous water heaters in caravans because supply systems have not usually been designed to allow for such heavy loading.

Each socket must have its own individual overcurrent protection in the form of a fuse or circuit breaker; to ensure that enough current will flow to open the protective device in the event of an earth fault, it is recommended that the earth-fault loop impedance at the plug should not exceed 2 Ω. All sockets must be protected by an RCD with a 30 mA rating, either individually or in groups, which must consist of no more than three sockets (*see* {Fig 7.12}). Since a 30 mA RCD is also required within each caravan, the device protecting the pitch socket(s) should be time delayed, to prevent the possibility of a fault in one caravan from switching off the supply to other caravans.

7.10 Highway power supplies and street furniture
7.10.1 Introduction ~ [611-01]

Regulations for highway power supplies feeding street furniture and street located equipment are new to the 16th Edition. The need for such regulations has arisen because of the increasing use of electrical supplies for street and footpath lighting, traffic signs, traffic control and traffic surveillance equipment. These uses are described as *street furniture*. Also covered by the Regulations is *street located equipment* which includes telephone kiosks, bus shelters, advertising signs and car park ticket dispensers. As far as the Regulations are concerned, no distinction is drawn between the two categories, both street furniture and street located equipment being subject to their requirements.

It must not be thought that these Regulations apply only to installations on public highways. They must also be followed in installations within private roads, car parks, and other areas where people will be present. They do not apply to *suppliers' works,* which in this context means the overhead or underground supplies feeding the equipment. It does apply, however, to overhead or underground supplies installed to interconnect street furniture or street located equipment from the point of supply.

7.10.2 Highway and street furniture regulations ~ [611-02 to 611-06]

The equipment covered by these regulations is always, by definition, accessible to people of all kinds; however, maintenance must only be carried out by skilled and/or instructed persons. Thus, certain measures for protection in areas only open to skilled or instructed persons are not appropriate here. For example, protection by barriers or by placing out of reach must not be used, because those using the areas may not be aware of the dangers which follow from climbing over barriers or reaching up to normally untouchable parts. The equipment considered here usually has doors which give access to the live parts inside. Such doors must not allow unauthorised access to live parts, and must therefore either:
1. be opened only with a special key or tool, or
2. be arranged so that opening the door disconnects the live parts, or
3. have a barrier inside the door (to give protection to IP2X - human fingers) to prevent contact with live parts when the door is open.

Many types of street furniture and street located equipment are fed using overhead cables. The usual protection against direct contact by 'placing out of reach' is acceptable for such systems, but special regulations apply to the equipment fed. Where uninsulated low voltage overhead conductors are more than 1.5 m vertically from the equipment, it may be maintained by a suitably *instructed person*. If the vertical clearance is less than 1.5 m, only a *skilled person* who has been trained in live working may be involved. All cables buried directly should have a marker tape placed above them, 150 mm below the ground surface. To prevent disturbance, burial depths are usually 450 mm below verges and 750 mm below the highway.

Protection against indirect contact (metalwork not normally expected to be live) cannot be achieved in these installations by a non-conducting location, earth-free equipotential bonding or electrical separation. The danger to the people who may touch metalwork is no more in this case than with electrical equipment indoors, so a 5 s disconnection time is acceptable where earthed equipotential bonding and automatic disconnection of the supply is employed. To prevent the effects of a fault being 'imported' or 'exported', adjacent metal structures should not be bonded to the circuit protection system.

Installations of this type are usually simple, often consisting of a single circuit feeding lighting. Provided that there are no more than two circuits, there is no need to provide a main switch or isolator, the supply cutout (main supply fuse in most cases) being used for this purpose, but only by instructed persons. However, where the supply is provided by a separate supplier (the Electricity Supply Company), their consent to use of the cutout for this purpose must first be obtained as with any other installation.

Internal wiring in all street electrical equipment must comply with the normal Regulations concerning protection, identification and support (*see* {4.5, 4.6 and 4.4.1}). Attention is drawn particularly to the need to support cables in vertical drops against undue stress. The notice indicating the need to provide periodic testing is unnecessary where an installation is subject to a planned inspection and test routine.

Temporary installations such as those for Christmas or summer external lighting schemes are usually connected to highway power supplies. In many cases, street furniture is equipped with *temporary supply units* from which such installations can be fed. The temporary supply unit must have a clear external label indicating the maximum current it is intended to supply. Attention is drawn to the possibility of damage to existing cable connections by the frequent connection and disconnection of temporary supplies. It is recommended that a socket outlet, especially intended to feed temporary installations, should be part of the temporary supply unit and should be fixed within the street furniture enclosure. Such installations must comply with the requirements for construction site installations (*see*{7.5}) and must not reduce the safety of the existing installation.

Highway equipment is usually subjected to vibration, corrosion and condensation, and sometimes to vandalism. It should be chosen with such problems in mind. The heat produced by lamps and control gear will usually be sufficient to prevent condensation and corrosion, but thought should be given to the provision of a low power heater in other cases.

Inspection and testing is necessary as for all electrical installations, and should be synchronised with other maintenance work, such as white lining and relamping, to avoid inconvenience to highway users as far as is possible. In general, a period of six years between tests is acceptable for fixed installations, and three months for temporary systems.

7.11   Heating appliances and installations

7.11.1   *Introduction* ~ [554-04]

This section is concerned with the special requirements for devices and circuits which are designed to produce heat for transfer to their surroundings. All water heaters (as well as those for other liquids) must be provided with a thermostat or cutout to prevent a dangerous rise in temperature. Not only could the high temperature of the water or other liquid be dangerous, but if allowed to boil, very high pressures, leading to the danger of an explosion, could result if the liquid container were sealed.

7.11.2   *Electrode boilers and water heaters* ~ [554-03]

An electrode heater or boiler is a device which heats the water contained, or raises steam.

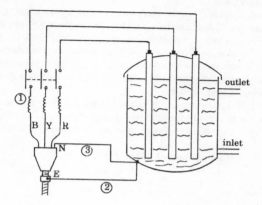

Fig 7.13   Three-phase electrode heater fed from a low voltage upply

Two or three electrodes are immersed in the water and a single- or three-phase supply is connected to them. There is no element, the water being heated by the current which flows through it between the electrodes. Electrode heaters and boilers must be used on a.c. supplies only, or electrolysis will occur, breaking down the water into its components of hydrogen and oxygen. The different requirements for single-phase and three-phase boilers are shown below.

A *Three-phase heaters fed from a low voltage supply {Fig 7.13}*
The requirements are:
1.   a controlling circuit breaker which opens all three phase conductors and is provided with protective overloads in each line,
2.   the shell of the heater bonded to the sheath and/or armour of the supply cable with a conductor of cross-sectional area at least equal to that of each phase conductor,
3.   the shell of the heater bonded to the neutral by a conductor of cross-sectional area at least equal to that of each phase conductor.
B *Three phase electrode heaters fed from a supply exceeding low voltage {Fig 7.14}*

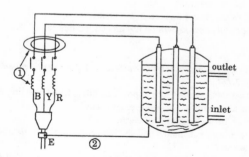

Fig 7.14   Three-phase electrode boiler fed from a supply exceeding low voltage

The requirements are:

1. a controlling circuit breaker fitted with residual current tripping, set to operate when the residual current is sustained and exceeds 10% of the rated supply current. Sometimes this arrangement will result in frequent tripping of the circuit breaker, in which case the tripping current may be reset to 15% of the rated current and/or a time delay device may be fitted to prevent tripping due to short duration transients. *See* {5.9} for more information concerning residual current devices,

2. the shell of the heater must be bonded to the sheath or armour of the supply cable with a conductor of current rating at least equal to the RCD tripping current, but with a minimum cross-sectional area of 2.5 mm^2.

C *Single-phase electrode heaters {Fig 7.15}*

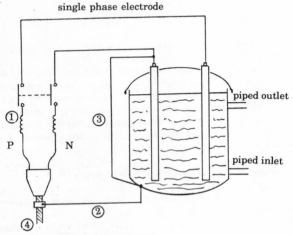

Fig 7.15   Single-phase electrode heater

The requirements for single-phase electrode heaters are:

1. a double-pole linked circuit breaker with overload protection in each line,
2. the shell of the heater bonded to the sheath and/or armour of the supply cable with a conductor of current-carrying capacity at least equal to that of each live conductor,
3. the shell of the heater bonded to the neutral,
4. the supply must be one which has an earthed neutral.

D *Insulated single-phase electrode heater, not permanently piped to water supply {Fig 7.16}*

This small heater is usually fixed in position, but is filled by using a flexible hose or a water container. The requirements are:

1. there must be no contact with earthed metal,
2. the heater must be insulated and shielded to prevent the electrodes from being touched,
3. control must be by means of a single-pole fuse or by a circuit breaker and a double pole switch,
4. the shell of the heater must be bonded to the cable sheath.

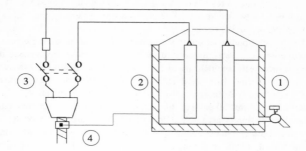

Fig 7.16   *Insulated single-phase electrode heater, not piped in to a*
             water supply

7.11.3   *Instantaneous water heaters*   ~   [554-05]

Water heaters of this type are in general use to provide hot water for showers, making drinks, and so on. They transfer heat from the element directly
to the water flowing over it, and therefore will be arranged to switch on only
when water is flowing. In most cases, the element of the heater has a fixed
rating, and so transfers energy in the form of heat to the water at a constant
rate. The temperature rise of the water passing over the element therefore
depends on the inlet water temperature and the rate of water flow. If the
discharge rate is high, the energy provided by the element may be insufficient to raise the water temperature to the desired level.

The slower the rate of flow, the hotter will become the water at the
outlet. This is the reason for the common complaint of being scalded whilst
under the shower if someone turns on a tap elsewhere, reducing the water
pressure and the rate of flow over the element. Some heaters are provided
with an automatic cutout to switch off the element if a preset outlet temperature is exceeded. {Figure 7.17} shows, in graphical form, the expected outlet water temperature for various water flow rates from 3 kW, 6 kW and
8 kW heaters assuming an inlet water temperature of 10°C.

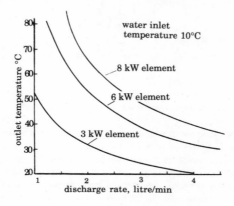

Fig 7.17   *Outlet temperatures from instantaneous water heater*

Some heaters are provided with thyristors or triacs to continuously vary the heater rating to maintain a desired water outlet temperature so long as the variation in the rate of flow is not too great. These devices adjust the effective current flow by delaying the instant in each half-cycle of the supply at which current begins to flow.

The Regulations point out that a heater with an uninsulated element is unsuitable where a water softener of the salt regenerative type is used because the increased conductivity of the water is likely to lead to excessive earth leakage currents from the usually uninsulated element. The agreement of the Water Supply Authority is usually needed before installation.

It is essential that all parts of this type of heater are solidly connected to the metal water supply pipe, which in turn is solidly earthed independently of the circuit protective conductor. The heater must be controlled by a double pole linked switch. In the case of a shower heater, if this switch is not built into the heater itself, a separate pull switch must be provided adjacent to the shower, with the switch itself being out of reach of a person using the shower. The arrangement of an instantaneous heater is shown in {Fig 7.18}. If the neutral supply to a heater with an uninsulated element is lost, current from the phase will return *via* the water and the earthed metal. Therefore, a careful check is necessary to ensure that there is no fuse, circuit breaker or non-linked switch in the neutral conductor. A problem frequently arises due to the increasing use of 3 kW instantaneous water heaters in caravans because supply systems have not usually been designed to allow for such heavy loading.

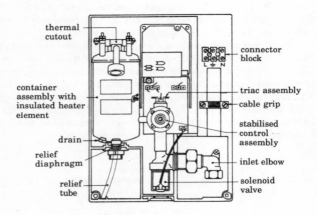

Fig 7.18   Instantaneous water heater

7.11.4   *Surface, floor, soil and road warming installations*   ~
[522-06-04, 554-06 and 554-07]

Most cables have conductors of very low resistance so that the passage of current through them dissipates as little heat as possible. By using a cable with a higher conductor resistance (typical resistances for some of the resistive alloys used are from 0.013 to 12.3 Ω/m) heat will be produced and will transfer to the medium in which the cable is buried. There are many types of cable, a typical example being shown in {Fig 7.19}.

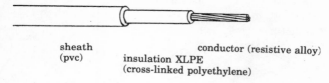

sheath   conductor (resistive alloy)
(pvc)   insulation XLPE
           (cross-linked polyethylene)

Fig 7.19   *One type of floor heating cable*

Some of the many examples of the use of heating cables are:
1. space heating, using the concrete floor slab as the storage medium,
2. under-pitch application on sports grounds, to keep the playing surfaces free of frost and snow,
3. in roads, ramps, pavements and steps to prevent icing,
4. surface heating cables, tapes and mats, used for frost protection, anti-condensation heating, process heating to allow chemical reactions, drying, processing thermoplastic and thermosetting materials, heating transport containers, *etc*
5. in rainwater drainage gutters to prevent blocking by ice and snow, and
6. for soil warming to promote plant growth in horticulture.

Table 7.7 Maximum conductor temperatures for floor warming cables
(from [Table 55C] of BS 7671: 1992)

Type of cable	Max conductor operating temperature (°C)
General purpose, p.v.c. over conductor	70
Enamelled conductor, polychloroprene ins. with p.v.c. sheath	70
Enamelled conductor, p.v.c. ins. overall	70
Enamelled conductor, p.v.c. ins. and lead alloy sheathed	70
Heat resisting p.v.c. insulated	85
Synthetic rubber insulated	85
Mineral insulated copper sheathed	*
Silicone treated woven glass insulation	180

* The operating temperature depends on the outer covering material, the type of seal, the arrangement of the cold tails, and so on. Manufacturers' data must be consulted

The heating cables used in such situations must be able to withstand possible damage from shovels, wheelbarrows, *etc.* during installation, as well as the corrosion and dampness which is likely to occur during use. They must be completely embedded, and installed so that they are not likely to suffer damage from cracking or movement in the embedding material, which is often concrete. The loading of the installation must be such that the temperatures specified for various types of conductor {Table 7.7} are not exceeded. Where an electrical under-floor heating system is used in a bath or shower room, it must either have an earthed metallic sheath which is supplementary bonded or be covered by an overall earthed metallic grid which is similarly bonded.

Where heating cables pass through, or run close to, materials which present a fire hazard, they must be protected from mechanical damage by a fire-proof enclosure. Where normal circuit cables are run through a heated floor, they must have the appropriate ambient temperature correction factor applied {4.3.4}. Heating cables may be obtained ready jointed to normal cables at their ends (cold tails) for connection to the supply circuit as shown in {Fig 7.20}.

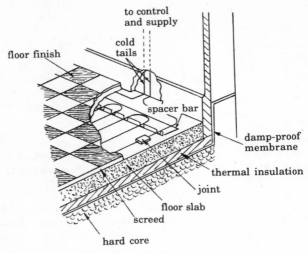

Fig 7.20   Floor heating installation

7.12   Discharge lighting

7.12.1   *Low voltage discharge lighting*

The very high luminous efficiency of discharge lamps has led to their almost universal application for industrial and commercial premises; the introduction of low rated types as direct replacements for filament lamps is beginning to see their wider use in domestic situations.

Discharge lamps are those which produce light as a result of a discharge in a gas. Included are:

Fluorescent

Really low pressure mercury vapour lamps, very widely used for general lighting in homes, shops, offices, *etc.*

High pressure mercury

Provide a very intense lighting level for outside use in situations where the (sometimes) poor colour rendering is not important.

Low pressure sodium

The most efficient lamp of all, but its poor colour (orange) light output limits its use to street and road lighting

High pressure sodium

The acceptable golden light colour enables the lamp to be used for road and outside lighting in areas where better colour rendering is needed, as well as for large indoor industrial applications.

Discharge lamps, unlike their incandescent counterparts, require control gear in the form of chokes, ballasts, auto-transformers and transformers. These devices result usually in a lagging power factor, which is corrected, at least partially, by connecting capacitance across the supply. This control gear should be positioned as close as possible to the lamps. Because of low power factor and the inductive/capacitive nature of the load, switches should be capable of breaking twice the rated current of a discharge lamp system, and maximum demand is calculated by using a multiplying factor of 1.8 {6.2.1}.

Electronic devices are becoming increasingly common to provide high voltage pulses to assist discharge lamps to strike (start). These pulses can cause problems with insulation breakdown in some types of cable, particularly low voltage mineral insulated types.

7.12.2 *High voltage discharge lighting* ~ [476-02-04, 476-03-05 to 476-03-07, 537-04-06 and 554-03]

Earlier versions of the 15th Edition of the Wiring Regulations included detailed requirements for high voltage discharge lighting circuits, which almost always means neon sign installations. The Regulations have now omitted most of this detail, which is contained in BS 559. Any electrician who will be concerned with the high voltage aspects of such installations should therefore take expert advice, which is likely to include making reference to the BS.

Low voltage wiring to feed the transformers in this kind of installation is similar to most other installations except that special requirements apply to the Fireman's Switch which controls the installation and allows the fire service to make sure that high voltages have been isolated before playing water on the system. These special requirements have been covered in {3.2.2} and will not be repeated here.

7.13 Underground and overhead wiring

7.13.1 *Overhead wiring types* ~ [471-13-03, 521-01-03, 522-08-05 and 522-08-06]

Most overhead wiring is carried out by the Electricity Supply Companies, and is not in the domain of the electrician. However, wiring between buildings is often necessary, and can be carried out using one or more of the following methods:

1. rubber or p.v.c. insulated cables bound to a separate catenary wire,
2. specialist cables with an integral catenary wire,
3. p.v.c. insulated and sheathed, or rubber insulated with an oil-resisting and flame retardant sheath, provided that precautions are taken to prevent chafing of the insulation or undue strain on the conductors,
4. bare or p.v.c. insulated conductors on insulators. In this case, the height of the cables and their positions must ensure that protection against direct contact is prevented,
5. cable sheathed and insulated as in 3) above enclosed in a single unjointed length of heavy-gauge steel conduit of at least 20 mm diameter.

Table 7.8 Maximum span and minimum height for overhead wiring between buildings

Type of system (see list in 7.13.1 for key)	Max. length of span (m)	Minimum height above ground		
		At road crossings (m)	Accessible to vehicles (m)	Not accessible to vehicles (m)
1	no limit	5.8	5.2	3.5
2	consult manufacturer	5.8	5.2	3.5
3	3	5.8	5.2	3.5
4 & 5	30	5.8	5.2	3.5
6	3	5.8	5.2	3.0

7.13.2 Maximum span lengths and minimum heights

The required information is contained in {Table 7.8} and {Fig 7.21}. Four special points are worth noting here.

1. in some locations where very tall traffic may be expected (*eg* yacht marinas, or where tall cranes are moved) heights above ground must be increased,
2. in agricultural situations, the lower heights of 3.5 m and 3.0 m are not permitted,

3. where caravan pitches are fed, overhead cables must be at least 2 m outside the bounds of each pitch,

4. overhead supplies to caravan pitches must have a minimum ground clearance of 6.0 m where there will be moving vehicles, and 3.5 m elsewhere.

NB All ground clearances are for positions
***NOT** accessible to vehicular traffic*

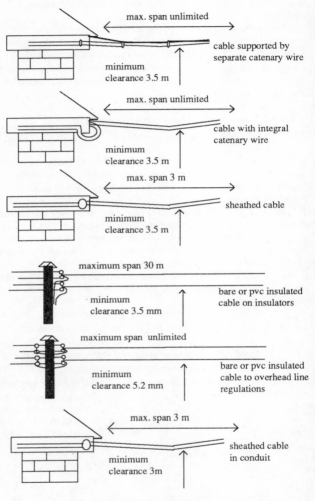

Fig 7.21   Overhead lines. Ground clearances shown are for places which are NOT accessible to vehicular traffic - see {Table 7.8}

7.13.3   Underground wiring ~ [522-04, 522-05 and 522-06-03]
Two types of cable may be installed underground:

1. armoured or metal sheathed or both

2. p.v.c. insulated concentric type. Such a cable will have the neutral (and possibly the protective conductor) surrounding the phase conductor. An example is shown in {Fig 4.1(c)}.

No specific requirements for depth of burial are given in the Regulations, except that the depth should be sufficient to prevent any disturbance of the ground reasonably likely to occur during normal use of the premises. Hence, a cable to outbuildings installed under a concrete path could be at 400 mm, whilst if running through a cultivated space which could be subject to double digging would be less likely to disturbance if buried at 700 mm. For caravan pitches, cables should be installed outside the area of the pitch, unless suitably protected, to avoid damage by tent pegs or ground anchors.

Cables must be identified by suitable tape or markers above the cable, so that anyone digging will become aware of the presence of the cable. Cable covers may also offer both identification and protection as shown in {Fig 7.22}.

It is often useful to lay a yellow cable marker tape just below ground level so that this will be exposed by digging before the cable is reached. Careful drawings should also be made to indicate the exact location of buried cables; such drawings will form part of the installation manual (*see* {8.2.1}).

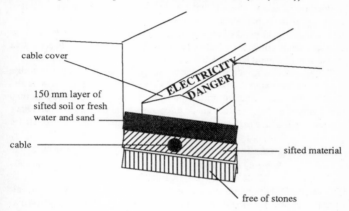

cable cover

150 mm layer of sifted soil or fresh water and sand

cable

sifted material

free of stones

Fig 7.22   Cable covers

7.14   Outdoor installations and garden buildings
7.14.1   *Temporary and garden buildings*
Many dwelling houses have buildings associated with them which are not directly part of the main structure. These include garages, greenhouses, summer houses, garden sheds, and so on. Many of them have an installation to provide for lighting and portable appliances. It is important to appreciate that the lightweight (and sometimes temporary) nature of such buildings does not reduce the required standards for the electrical installation. On the contrary, the standards of both the installation and its maintenance may need to be higher to allow for the arduous conditions.

Particular attention must be paid to the following:

1. supplies to such outbuildings must comply with the requirements for overhead and underground supplies stated in {7.13},
2. the equipment selected and installed must be suitable for the environment in which it is situated. For example, a heater for use in a greenhouse will probably meet levels of humidity, temperature and spraying water not encountered indoors, and should be of a suitably protected type,
3. the earthing and bonding must be of the highest quality because of the increased danger in outdoor situations. All socket outlets should be protected by 30 mA RCDs.

7.14.2 Garden installations ~ [471-16]

Increasing use is being made of electrical supplies in the garden, for pond pumping sytems, lighting, power tools and so on. The following points apply:

1. Socket outlets installed indoors but intended to provide outdoor supplies must be protected by an RCD with a maximum operating current of 30 mA.

2. Garden lighting, pond pumps and so on should preferably be of Class III construction, supplied from a SELV system and having a safety isolating transformer supply. Where 240 V equipment must be used, it should be Class II double insulated (no earth) and should be suitably protected against the ingress of dust or water.

3. Earthing must be given special attention. All buildings must be provided with 30 mA RCD protection, but the Electricity Supply Company should be consulted to ascertain their special requirements if the supply system uses the PME (TN-C-S) system. Where the supplier does not provide an earth terminal, each outbuilding must be provided with an adjacent earth electrode.

4. Outbuildings are often of light construction and therefore are subject to extremes as far as temperature swings are concerned. It is therefore important to bear this in mind when selecting equipment and components.

5. Extraneous conductive parts of an outbuilding which may become live due to a fault should be bonded to the incoming protective conductor.

6. Every outbuilding with an electrical supply should be provided with a means of isolation to disconnect all live conductors including the neutral.

7. All outbuildings where protection against direct contact is by earthed equipotential bonding and automatic disconnection of the supply should have a disconnection time in the event of an earth fault which does not exceed 0.4 s.

7.15 Installation of machines and transformers

7.15.1 Rotating machines ~ [130-06-02, 473-03-01 and 473-03-02, 476-02, 476-03-01 to 476-03-04 and 552]

The vast majority of motors used in industry are of the three-phase squirrel-cage induction type. Smaller motors are usually single-phase induction machines. Induction motors have important advantages, such as robustness, minimal maintenance needs, and self starting characteristics, but all draw very high starting currents from their supplies {Fig 7.23}. This starting current is a short-lived transient, and may usually be ignored when calculating cable sizes.

Although the starting current may be several times the running current, the value depending on the machine characteristics and the connected mechanical load, its short duration will not lead to overheating in usual circumstances. If frequent starting is a requirement, larger supply cables may be necessary to avoid damage to insulation. A problem could arise when fast-acting fuses or circuit breakers are used for short-circuit protection; the high starting current may result in operation of the protective device. A common, but unsatisfactory, remedy for this difficulty is to increase the rating of the protective device, leading to a loss of proper overload protection. A possible solution is to use dual rated fuses (gM types). For example, a 25M40 fuse has a continuous rating of 25 A and the operating characteristics of a 40 A fuse.

A word is necessary concerning motor ratings. Many years ago it was decided to replace the *horsepower* as the unit of output power with the kilowatt. Unfortunately, the old horsepower is a very long time in dying. Many machines still have rating plates giving output power in horsepower. The conversion is straightforward. Since one horsepower is the same as 746 W, horsepower is converted to kilowatts by multiplying by 0.746.

It is sometimes practice to stop a motor very quickly by feeding it with a reverse current. When this method is provided it is important that the machine does not begin to move in the reverse direction if this would cause danger.

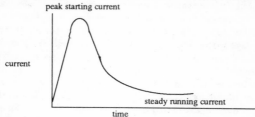

Fig 7.23   *Starting current of an induction motor*

Where other types of motor, such as wound rotor and commutator induction or thyristor fed dc types are used, the cables must be suitable for carrying running currents on full load, which will usually mean that they are large enough to carry the short duration starting currents.

Every motor rated at 0.37 kW (0.5 horsepower) or more must be fed from a starter which includes overload protection. Such devices have time-delay features so that they will not trip as a result of high starting current, but will do so in the event of a small but prolonged overload. They have the advantages over fuses and single-pole circuit breakers that all three lines of a three-phase system are tripped by an overload in any one of them. If only one line were broken, the resulting 'single-phasing' operation of the motor could cause it to overheat.

It is often necessary to provide a means to prevent automatic restarting after failure of the supply. For example, if the supply to a machine shop fails, the machine operators are likely to use the enforced break in production to clean and service their machines. If so, when the supply is restored, the presence of hands, brushes, tools, *etc.* in the machines when they automatically restart would cause serious danger. The necessary 'no-volt protection' is obtained by using a starter of the type whose circuit is shown in {Fig 7.24}. The coil is fed through the 'hold-in' contacts, which open when the supply fails; the motor can then be operated only by pressing the 'start' button. This requirement does not apply to protected motors which are required to restart automatically after mains failure. Examples are motors supplying refrigeration and pumping plants. It is important that lock-off stop buttons are not used as a means of isolation.

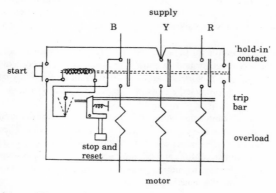

Fig 7.24   *Direct-on-line starter*

7.15.2 *Transformers* ~ [551]

Transformers are used more widely in the supply system than in the installation itself. The link between the primary and secondary windings of a double wound transformer is magnetic, not electric, so there is no electrical connection with the supply system, or with its earthing system, from a circuit fed by the secondary winding of a transformer. This loss of earthing may be an advantage where electrical separation is required, for example, where an electric shaver in a bathroom is fed from the secondary winding of a transformer (*see* {5.8.4}).

Step-up transformers must not be used in IT systems. In systems where they are permitted, linked multipole switches must be provided so that the supply is simultaneously disconnected from all live conductors, including the neutral.

7.16 Reduced voltage systems

7.16.1 *Types of reduced voltage* ~ [411-01]

Most installations operate at low voltage, which is defined as up to 1000 V ac or 1500 V dc between conductors, and up to 600 V ac or 900 V dc between conductors and earth. Extra-low voltage is defined as not exceeding 50 V ac or 120 V dc between conductors or from conductors to earth. Four types of reduced voltage system, all intended to improve safety, are recognised by the Regulations. They are:

1. separated extra-low voltage or SELV {7.16.2},
2. functional extra-low voltage or FELV {7.16.3},
3. reduced voltage {7.16.4},
4. voltages at values lower than the maximum for ELV may be required to ensure the safety of people and of livestock in areas where body resistance is likely to be very low {7.6}.

7.16.2 *Separated extra-low voltage (SELV)* ~ [411-02 and 471-16]

The safety of this system stems from its low voltage level, which should never exceed 50 V ac or 120 V dc, and is too low to cause enough current to flow to provide a lethal electric shock. The reason for the difference between ac and dc levels is shown in {Figs 3.9 & 3.10}.

It is *not* intended that people should make contact with conductors at this voltage; where live parts are not insulated or otherwise protected, they must be fed at the lower voltage level of 25 V ac or 60 V ripple-free dc although insulation may sometimes be necessary, for example to prevent short-circuits on high power batteries. To qualify as a separated extra-low voltage (SELV) system, an installation must comply with conditions which include:

1. it must be impossible for the extra-low voltage source to come into contact with a low voltage system. It can be obtained from a safety isolating transformer, a suitable motor generator set, a battery, or an electronic power supply unit which is protected against the appearance of low voltage at its terminals,

2. there must be no connection whatever between the live parts of the SELV system and earth or the protective system of low voltage circuits. The danger here is that the earthed metalwork of another system may rise to a high potential under fault conditions and be imported into the SELV system.

3. there must be physical separation from the conductors of other systems, the segregation being the same as that required for circuits of different types {6.6},

4. plugs and sockets must not be interchangeable with those of other systems; this requirement will prevent a SELV device being accidentally connected to a low voltage system,

5. plugs and sockets must *not* have a protective connection (earth pin). This will prevent the mixing of SELV and FELV devices. Where the Electricity at Work Regulations 1989 apply, sockets *must* have an earth connection, so in this case appliances must be double insulated to class II so that they are fed by a two-core connection and no earth is required,

6. luminaire support couplers with earthing provision must not be used.

7.16.3   *Functional extra-low voltage (FELV ~ [411-03 & 471-14]*

If a separated extra-low voltage (SELV) system is earthed, or if the insulation of the supply which feeds it does not meet the necessary requirements, it ceases to be a SELV system and becomes a functional extra-low voltage (FELV) system. This difference is illustrated by {Fig 7.25}.

If a functional system only fails to be classified as a safety system because it is earthed, it must then be protected by enclosures that prevent it being touched or by insulation capable of withstanding a test voltage of 500 V r.m.s. for one minute. In such a case there is no need for the earth-fault loop impedance of the extra-low voltage circuit to be low enough to prevent danger; compliance of the safety source supply is sufficient. There is therefore no need to bond and earth all associated non-current-carrying metalwork.

Such metalwork must, however, be bonded and earthed if the insulation between the low-voltage supply circuit and the functional extra-low voltage circuit does not meet the requirements stated above. In the event of the supply circuit being of the earth-free bonded type {5.8.3} all non-current-carrying metalwork of the FELV system must be connected to the non-earthed protective conductor of the supply circuit. Plugs and sockets in FELV systems must not be interchangeable with those of other supply systems in use in the same premises.

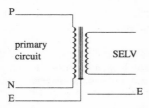

becomes a fonctional extra-low voltage system when:

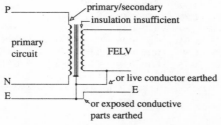

Fig 7.25   *Earthing relationship, SELV and FELV systems*

7.16.4   *Reduced voltage* ~ [471-15]

In situations where the power requirements are high, extra-low voltage systems would need to deliver very high currents. If an increased voltage can still lead to a safe system, the current required will be reduced. We are therefore considering a voltage higher than extra-low but lower than low voltage.

The highest voltage permitted for such systems is 110 V between conductors for both single- and three-phase systems, with voltages to earth of 55 V for single-phase and 65 V for three-phase supplies {Fig 7.26}.

The reduced voltage supply can be taken from a double wound isolating transformer, or from a suitable motor generator set, provided it has the unusual feature of a centre-tapped winding. The system must be insulated and protected against direct contact as for a low voltage installation. Earth-fault loop impedance must allow automatic disconnection in a maximum of 5 s (*see* {Table 7.9}), or an RCD with an operating current of no more than 30 mA must be provided. The result of multiplying earth-fault loop impedance (Ω) by the RCD operating current (A) must not exceed 50 (V). Plugs, sockets and cable couplers in the reduced voltage system must all be provided with protective conductor contacts (earth pins) and must not be interchangeable with those of any other system in use at the same location.

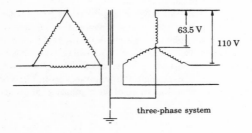

three-phase system

Fig 7.26   Reduced voltage systems

Table 7.9   Maximum earth-fault loop impedance (Ω) for 5 s disconnection time in reduced voltage systems
(from [Table 471A] of BS 7671: 1992)

Device rating (A)	MCB type 1 55 V	MCB type 1 63.5 V	MCB type 2 55 V	MCB type 2 63.5V	MCB types C & 3 55 V	MCB types C & 3 63.5 V	BS 88 fuse 55 V	BS 88 fuse 63.5V
6	2.30	2.65	1.32	1.52	0.92	1.07	3.20	3.70
10	1.38	1.59	0.79	0.91	0.55	0.64	1.77	2.05
16	0.86	0.99	0.49	0.57	0.34	0.40	1.00	1.15
20	0.69	0.80	0.40	0.46	0.28	0.32	0.69	0.80
25	0.55	0.64	0.32	0.36	0.22	0.26	0.55	0.63
32	0.43	0.50	0.25	0.28	0.17	0.20	0.44	0.51
50	0.28	0.32	0.16	0.18	0.11	0.13	0.25	0.29

Note: 63.5 V is the voltage to earth (to the star point) of a 110 V three-phase supply.

Inspection and testing

8.1 Introduction

8.1.1 The tester

The person who carries out the test and inspection must be competent to do so, and must be able to ensure his own safety, as well as that of others in the vicinity. It follows that he must be skilled and have experience of the type of installation to be inspected and tested so that there will be no accidents during the process to people, to livestock, or to property. The Regulations do not define the term 'competent', but it should be taken to mean a qualified electrician or electrical engineer.

8.1.2 Why do we need inspection and testing? ~ [711-01 and 713-01-01]

There is little point in setting up Regulations to control the way in which electrical installations are designed and installed if it is not verified that they have been followed. For example, the protection of installation users against the danger of fatal electric shock due to indirect contact is usually the low impedance of the earth-fault loop; unless this impedance is correctly measured, this safety cannot be confirmed. In this case the test cannot be carried out during installation, because part of the loop is made up of the supply system which is not connected until work is complete.

In the event of an open circuit in a protective conductor, the whole of the earthed system could become live during the earth-fault loop test. The correct sequence of testing {8.3} would prevent such a danger, but the tester must always be aware of the hazards applying to himself and to others due to his activities. Testing routines must take account of the dangers and be arranged to prevent them. Prominent notices should be displayed to indicate that no attempt should be made to use the installation whilst testing is in progress.

The precautions to be taken by the tester should include the following:

1. make sure that all safety precautions are observed
2. have a clear understanding of the installation, how it is designed and how it has been installed
3. make sure that the instruments to be used for the tests are to the necessary standards (BS 4743 and BS 5458) and have been recently recalibrated to ensure their accuracy
4. check that the test leads to be used are in good order, with no cracked or broken insulation or connectors, and are fused where necessary to comply with the Health and Safety Executive Guidance Note GS38
5. be aware of the dangers associated with the use of high voltages for insulation testing. For example, cables or capacitors connected in a circuit which has been insulation tested may have become charged to a high potential and may hold it for a significant time.

The dangers associated with earth-fault loop impedance testing have been mentioned in {5.3}.

8.1.3 *Information needed by the tester* [514-08 and 514-09, 712-01]

The installation tester, as well as the user, must be provided with clear indications as to how the installation will carry out its intended purpose.

Table 8.1 Typical schedule of circuits
(refer to {Fig 8.1})

Circuit	Fuse (BS 88)	Cable	Feeding
A1	32A	2.5mm² p.v.c. (X)	ring circuit, printing section
A2	32A	2.5mm² p.v.c. (X)	ring circuit, drawing office sockets
A3	32A	2.5mm² p.v.c. (X)	ring circuit, main office sockets
A4	32A	4.0mm² p.v.c. (Z)	store room printer
A5	spare		
A6	spare		
B1	10A	1.5mm² p.v.c. (Y)	lighting, printing section
B2	10A	1.5mm² p.v.c. (Y)	lighting, drawing office
B3	10A	1.5mm² p.v.c. (Y)	lighting, main office
B4	10A	1.5mm² p.v.c. (Y)	lighting, corridors and toilets
B5	spare		
B6	spare		
C1	32A	2.5mm² p.v.c. (X)	ring circuit, paper store
C2	32A	2.5mm² p.v.c. (X)	ring circuit, binding section
C3	spare		
C4	spare		
C5	10A	1.5mm² p.v.c. (Y)	lighting, paper store
C6	10A	1.5mm² p.v.c. (Y)	lighting, binding section

To this end, the person carrying out the testing and inspection must be provided with the following data:

1 the type of supply to be connected, *ie* single- or three- phase
2 the assessed maximum demand {6.2.1}
3 the earthing arrangements for the installation
4 full details of the installation design, including the number and position of mains gear and of circuits, (*see* {Table 8.1},{Fig 8.1} and {Table 8.2} for a typical example)
5 all data concerning installation design, including calculation of live and protective conductor sizes, maximum demand, *etc.*
6 the method chosen to prevent electric shock in the event of an earth fault.

Without this complete information the tester cannot verify either that the installation will comply with the Regulations, or that it has been installed in full accordance with the design.

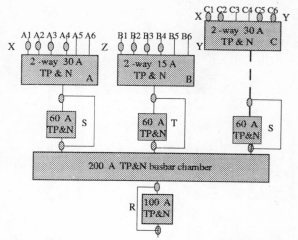

Fig 8.1   Typical mains gear schematic diagram. See also {Tables 8.1 &8.2}

8.2   Inspection

8.2.1   *Notices and other identification*   ~   [514-10 to 514-12]

The installation tester, as well as the user, must have no difficulty in identifying circuits, fuses, circuit breakers, *etc*. He must make sure that the installation is properly equipped with labels and notices, which should include:

1   Labels for all fuses and circuit breakers to indicate their ratings and the circuits protected

2   Indication of the purpose of main switches and isolators

3   A diagram or chart at the mains position showing the number of points and the size and type of cables for each circuit, the method of providing protection from direct contact and details of any circuit in which there is equipment such as passive infra-red detectors or electronic fluorescent starters vulnerable to the high voltage used for insulation testing.

4   Warning of the presence of voltages exceeding 250 V on an equipment or enclosure where such a voltage would not normally be expected.

5   Warning that voltage exceeding 250 V is present between separate pieces of equipment which are within arm's reach

6   A notice situated at the main intake position to draw attention to the need for periodic testing (*see* {Table 8.12})

7   A warning of the danger of disconnecting earth wires at the point of connection of:

a)   the earthing conductor to the earth electrode

b)   the main earth terminal, where separate from main switchgear

c)   bonding conductors to extraneous conductive parts
The notice should read

Safety electrical connection — do not remove

8   A notice to indicate the need for periodic testing of an RCD as indicated in {5.9.2}

9   A notice for caravans so as to draw attention to the connection and dis connection procedure as indicated in {Table 7.5}

10   Warning of the need for operation of two isolation devices to make a piece of equipment safe to work on where this applies

11   A schedule at each distribution board listing the items to be disconnected (such as semiconductors) so that they will not be damaged by testing.

12   A drawing which shows clearly the exact position of all runs of buried cables.

Table 8.2 Typical mains gear schedule
(refer to {Fig 8.1})

Item	Situation	Supplying and type	Cable size BS 88	Fuse size
Dist.board A 2-way 30A TP&N	Switch room sockets	Ground floor	16mm² p.v.c (S)	4 x 32 A (2 spare ways)
Dist. board B 2-way 15A TP&N	Switch room	Ground floor lighting	6mm² p.v.c. (T)	4 x 10 A (2 spare ways)
Dist. board C 2-way 30A TP&N	Upper store	Upper floor installation	16mm² p.v.c. (S)	2 x 32 A 2 x 10 A (2 spare ways)
Board A switch-fuse 60A TP&N	Switch room	Dist. board A	16mm² p.v.c. (S)	3 x 60 A
Board B switch-fuse 30A TP&N	Switch room	Dist. board B	6mm² p.v.c. (T)	3 x 30 A
Board C switch-fuse 60A TP&N	Switch room	Dist. board C	16mm² p.v.c. (S)	3 x 60A
Main fuse-switch 100A TP&N	Switch room	Complete installation	70mm² p.v.c. (R)	3 x 100A

8.2.2   *Inspection* ~ [712-01 and 742]

Before testing begins it is important that a full inspection of the complete installation is carried out with the supply disconnected. The word 'inspection' has replaced 'visual inspection', indicating that all the senses (touch, hearing and smell, as well as sight) must be used. The main purpose of the inspection is to confirm that the equipment and materials installed:

1   are not obviously damaged or defective so that safety is reduced
2   have been correctly selected and erected
3   comply with the applicable British Standard or the acceptable equivalent
4   are suitable for the prevailing environmental conditions

An inspection check-list is shown in Table 8.3.

Some inspections are best carried out whilst the work is in progress. A good example is the presence of fire barriers within trunking or around conduit where they pass through walls.

Table 8.3 Inspection check list
(from [712-01-03] of BS 7671: 1992)

1	Identification of conductors
2	Mechanical protection for cables, or routing in safe zones
3	Connection of conductors
4	Correct connection of lampholders, socket outlets, *etc.*
5	Connection of single-pole switches in phase conductors only
6	Checking of design calculations to ensure that correct live and protective conductors have been selected in terms of their current-carrying capacity and volt drop
7	Presence of fire barriers
8	Protection of live parts by insulation to prevent direct contact
9	Protection against indirect contact by the use of:-
	a) protective conductors
	b) earthing conductors
	c) main and supplementary equipotential bonding conductors
10	Prevention of mutual detrimental influence
11	Electrical separation
12	Undervoltage protective devices
13	Use of Class II equipment
14	Labelling of fuses, circuit breakers, circuits, switches and terminals
15	The settings and ratings of devices for protection against indirect contact and against overcurrent
16	The presence of diagrams, instructions, notices, warnings, *etc.*
17	Selection of protective measures and equipment in the light of the external influences involved (such as the presence of disabled people)

8.2.3 *Periodic inspection* ~ [514-12-02, 731-01, 732-01 and 744-01]

No electrical installation, no matter how carefully designed and erected, can be expected to last forever. Deterioration will take place due to age as well as due to normal wear and tear. With this in mind, the Regulations require regular inspection and testing to take place so that the installation can be maintained in a good and a safe condition. It is now a requirement of the Regulations that the installation user should be informed of the need for periodic testing, and the date on which the next test is due.

Accessories, switchgear *etc* should be carefully examined for signs of overheating. Structural changes may have impaired the safety of an installation, as may have changes in the use of space. The use of extension leads must be discouraged, if only because of the relatively high loop impedance they introduce.

It is important to appreciate that the regular inspection and testing of all electrical installations is a requirement of the Electricity at Work Regulations. The time interval concerned will, of course, depend on the type of installation and on the way in which it is used. {Table 8.4} shows the suggested intervals between periodic tests and inspections.

8.3   Testing sequence

8.3.1   *Why is correct sequence important?*

Testing can be hazardous, both to the tester and to others who are within the area of the installation during the test. The danger is compounded if tests are not carried out in the correct sequence.

For example, it is of great importance that the continuity, and hence the effectiveness, of protective conductors is confirmed before the insulation resistance test is carried out. The high voltage used for insulation testing could appear on all extraneous metalwork associated with the installation in the event of an open-circuit protective conductor if insulation resistance is very low.

Table 8.4   Suggested intervals between periodic tests and inspections

Type of Installation	*Maximum period between inspections*
Domestic premises	10 years
Commercial premises	5 years
Educational establishmehts	5 years
Hospitals	5 years
Industrial premises	3 years
Cinemas	1 year*
Churches under five years old	2 years
Churches over five years old	1 year
Leisure complexes	1 year
Places of public entertainment	1 year
Theatres, *etc.*	1 year*
Agricultural and horticultural	3 years
Caravans	3 years
Caravan sites	1 year*
Emergency lighting	3 years
Fire alarm systems	1 year
Launderettes	1 year*
Petrol filling stations	1 year*
Highway power supplies	6 years
Temporary installations	3 months

Where maximum periods are marked * there is a legal requirement for retests at these intervals

Again, an earth-fault loop impedance test cannot be conducted before an installation is connected to the supply, and the danger associated with such a connection before verifying polarity, protective system effectiveness and insulation resistance will be obvious.

8.3.2   *Correct testing sequence* ~ [713-01]
Some tests will be carried out before the supply is connected, whilst others cannot be performed until the installation is energised. {Table 8.5} shows the correct sequence of testing to reduce the possibility of accidents to the minimum.

Table 8.5   Correct sequence for safe testing
(from [713] of BS 7671: 1992)

BEFORE CONNECTION OF THE SUPPLY

1	Continuity of protective conductors
2	Main and supplementary bonding continuity
3	Continuity of ring final circuit conductors
4	Insulation resistance
5	Site applied insulation
6	Protection by separation
7	Protection by barriers and enclosures
8	Insulation of non-conducting floors and walls
9	Polarity

WITH THE SUPPLY CONNECTED

10	Earth electrode resistance
11	Confirm correct polarity
12	Earth-fault loop impedance
13	Correct operation of residual current devices
14	Correct operation of switches and isolators

8.4 Continuity tests

8.4.1 *Protective conductor continuity* ~ [413-4, 713-2]

All protective and bonding conductors must be tested to ensure that they are electrically safe and correctly connected. {8.7.1} gives test instrument requirements. Provided that the supply is not yet connected, it is permissible to disconnect the protective and equipotential conductors from the main earthing terminal to carry out testing. Where the mains supply is connected, as will be the case for periodic testing, the protective and equipotential conductors *must not* be disconnected because if a fault occurs these conductors may rise to a high potential above earth. In this case, an earth-fault loop tester can be used to verify the integrity of the protective system.

Where earth-fault loop impedance measurement of the installation is carried out, this will remove the need for protective conductor tests because that conductor forms part of the loop. However, the loop test cannot be carried out until the supply is connected, so testing of the protective system is necessary before supply connection, because connection of the supply to an installation with a faulty protective system could lead to danger.

There are three methods for measurement of the resistance of the protective conductor.

1 *Using the neutral conductor as a return lead*

A temporary link is made at the distribution board between neutral and protective conductor systems. DON'T FORGET TO REMOVE THE LINK AFTER TESTING. The low resistance tester is then connected to the earth and neutral of the point from which the measurement is taken (*see* {Fig 8.2}). This gives the combined resistance of the protective and neutral conductors back to the distribution board. Then

$$R_p = R \times \frac{A_n}{A_n + A_p}$$

where

R_p is the resistance of the protective conductor

R is the resistance reading taken

A_n is the cross-sectional area of the neutral conductor

A_p is the cross-sectional area of the protective conductor.

Note that the instrument reading taken in this case is the value of the resistance $R_1 + R_2$ calculated from {Table 5.5} (*see* {8.4.4}).

2 *Using a long return lead*

This time a long lead is used which will stretch from the main earthing terminal to every point of the installation.

First, connect the two ends of this lead to the instrument to measure its resistance. Make a note of the value, and then connect one end of the lead to the main earthing terminal and the other end to one of the meter terminals.

Second, take the meter with its long lead still connected to the point from which continuity measurement is required, and connect the second meter terminal to the protective conductor at that point.

The reading then taken will be the combined resistance of the long lead and the protective conductor, so the protective conductor value can be found by subtracting the lead resistance from the reading.

$$R_p = R - R_L$$

where R_p is the resistance of the protective conductor

R is the resistance reading taken

R_L is the resistance of the long lead.

Some modern electronic resistance meters have a facility for storing the lead resistance at the touch of a button, and for subtracting it at a further touch.

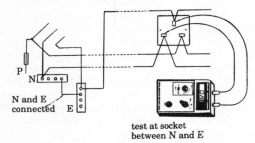

*test at socket
between N and E*

*Fig 8.2   Protective conductor continuity test using the neutral
conductor as the return lead*

3   *Where ferrous material forms all or part of the protective
           conductor*

There are some cases where the protective conductor is made up wholly or in part by conduit, trunking, steel wire armour, and the like. The resistance of such materials will always be likely to rise with age due to loose joints and the effects of corrosion. Three tests may be carried out, those listed being of increasing severity as far as the current-carrying capacity of the protective conductor is concerned. They are:

1   A standard ohmmeter test as indicated in *1* or *2* above. This is a low current test which may not show up poor contact effects in the conductor. Following this test, the conductor should be inspected along its length to note if there are any obvious points where problems could occur.

2   If it is felt by the inspector that there may be reasons to question the soundness of the protective conductor, a phase-earth loop impedance test should be carried out with the conductor in question forming part of the loop. This type of test is explained more fully in {8.4.4}

3   If it is still felt that the protective conductor resistance is suspect, the high current test using 1.5 times the circuit design current (with a maximum of 25 A) may be used. The circuit arrangement for such a test is shown in {Fig 8.3}. The protective circuit resistance together with that of the wander lead can be calculated from:

$$\frac{\text{voltmeter reading (V)}}{\text{ammeter reading (A)}}$$

Subtracting wander lead resistance from the calculated value will give the resistance of the protective system.

The resistance between any extraneous conductive part and the main earthing terminal should be 0.05 Ω or less; all supplementary bonds are also required to have the same resistance.

8.4.2   *Ring final circuit continuity*   ~   [713-03]

The ring final circuit, feeding 13 A sockets, is extremely widely used, both in domestic and in commercial or industrial situations. It is very important that each of the three rings associated with each circuit (phase, neutral and protective conductors) should be continuous and not broken. If this happens, current will not be properly shared by the circuit conductors. {Fig 8.4} shows how this will happen. {Fig 8.4(a)} shows a ring circuit feeding ten socket outlets, each of which is assumed to supply a load taking a current of 3 A. In simple terms, current is then shared between the conductors, so that each could have a current carrying capacity of 15 A. {Fig 8.4(b)} shows the same ring circuit with the same loads, but broken between the ninth and tenth sockets. It can be seen that now one cable will carry only 3 A whilst the other (perhaps with a current rating of 15 A) will carry 27 A. The effect will occur in any broken ring, whether simply one live conductor or both are broken.

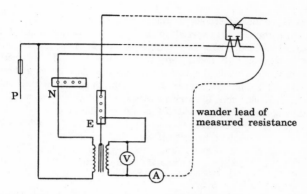

Fig 8.3   High-current ac test of a protective conductor

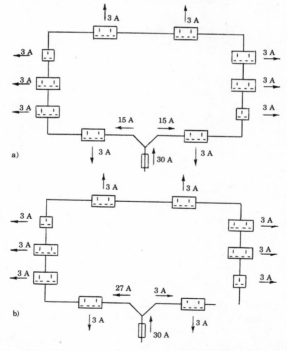

Fig 8.4   Illustrating the danger of a break in a ring final circuit
a) unbroken ring with correct current sharing
b) broken ring with incorrect current sharing

It is similarly important that there should be no 'bridge' connection across the circuit. This would happen if, for example, two spurs from different points of the ring were connected together as shown in [Fig 8.5], and again could result in incorrect load sharing between the ring conductors.

The tests of the ring final circuit will establish that neither a broken nor a bridged ring has occurred. The following suggested test is based on the Guidance Note on Inspection and Testing issued by the IEE.

Test 1

This test confirms that complete rings exist and that there are no breaks. To complete the test, the two ends of the ring cable are disconnected at the distribution board. The phase conductor of one side of the ring and the neutral from the other (P_1 and N_2) are connected together, and a low resistance ohmmeter used to measure the resistance between the remaining phase and the neutral (P_2 and N_1). {Figure 8.6} shows that this confirms the continuity of the live conductors. To check the continuity of the circuit protective conductor, connect the phase and CPC of different sides together (P_1 and E_2) and measure the resistance between phase and CPC of the other side (P_2 and E_1). The result of this test will be a measurement of the resistance of live and protective conductors round the ring, and if divided by four gives ($R_1 + R_2$) which will conform with the values calculated from {Table 5.5}.

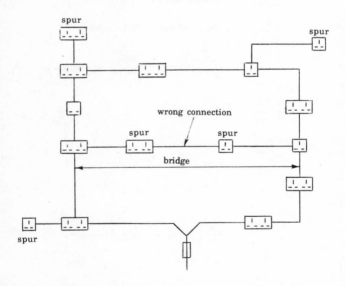

Fig 8.5   A 'bridged' ring final circuit

Test 2

This test will confirm the absence of bridges in the ring circuit. First, the phase conductor of one side of the ring is connected to the neutral of the other (P_1 and N_2) and the remaining phase and neutral are also connected together (P_2 and N_1). The resistance is then measured between phase and neutral contacts of each socket on the ring. If the results of these measurements are all substantially the same, the absence of a bridge is confirmed. If the readings are different, this will indicate the presence of a bridge or may be due to incorrect connection of the ends of the ring. If they are connected

P_1 to N_1 and P_2 to N_2 then readings will increase or reduce as successive measurements round the ring are taken, as is the case where a bridge exists. Whilst this misconnection is easily avoided when using sheathed cables, a mistake can be made very easily if the system consists of single-core cables in conduit. It may be of interest to note that the resistance reading between phase and neutral outlets at each socket should be one quarter of the phase/neutral reading of Test 1.

Measurements are also taken at each socket on the ring between the phase and the protective conductor with the temporary connection made at the origin of the ring between P_1 to E_2 and between P_2 to E_1. Substantially similar results will indicate the absence of bridges.

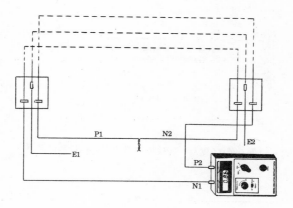

Fig 8.6   Test to confirm the continuity of a ring final circuit

8.4.3   *Correct polarity*   ~   [713-09]

If a single-pole switch or fuse is connected in the neutral of the system rather than in the phase, a very dangerous situation may result as illustrated in {Fig 8.8}.

It is thus of the greatest importance that single-pole switches, fuses and circuit breakers are connected in the phase (non-earthed) conductor, and verification of this connection is the purpose of the polarity test. Also of importance is to test that the outer (screw) connection of E S lampholders is connected to the earthed (neutral) conductor, as well as the outer contact of single contact bayonet cap (BC) lampholders. The test may be carried out with a long wander lead connected to the phase conductors at the distribution board and to one terminal of an ohmmeter or a continuity tester on its low resistance scale. The other connection of the device is equipped with a shorter lead which is connected in turn to switches, centre lampholder contacts, phase sockets of socket outlets and so on. A very low resistance reading indicates correct polarity (*see* {Fig 8.9}).

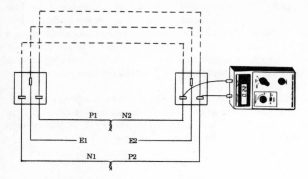

Fig 8.7   Test to confirm the absence of bridges in a ring final circuit

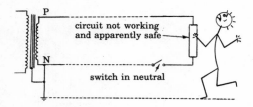

Fig 8.8   The danger of breaking the neutral of a circuit

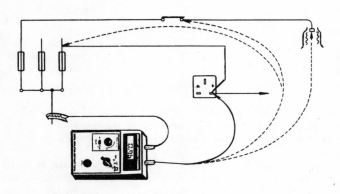

Fig 8.9   Polarity test of an installation

To avoid the use of a long test lead, a temporary connection of phase to
protective systems may be made at the mains position. A simple resistance
test between phase and protective connections at each outlet will then verify
polarity. In the unlikely event of the phase and protective conductor connections having been transposed at the outlet, correct polarity will still be shown
by this method; this error must be overcome by visual verification
    DON'T FORGET TO REMOVE THE TEMPORARY CONNECTION
                                    AFTERWARDS!

Special care in checking polarity is necessary with periodic tests of installations already connected to the supply, which must be switched off before polarity testing. It is also necessary to confirm correct connection of supply phase and neutral. Should they be transposed, all correctly-connected single-pole devices will be in the neutral, and not in the phase conductor.

One practical method of checking polarity and continuity of ring or radial circuits for socket outlets is to connect two low power lamps to a 13 A plug. One is connected between phase and neutral, and the other between phase and earth. Plugging in at each socket tests correct polarity and the continuity of live and protective conductors when both indicators light. It is important where RCD protection is employed to use very low power indicator lamps such as neon or LED devices (with suitable current limiting resistors where necessary). The smallest filament lamp will take sufficient current from phase to earth to trip most RCDs.

8.4.4 Measurement of $R_1 + R_2$

In {5.3.6} and in {Table 5.5} we considered the value of the resistance of the phase conductor plus that of the protective conductor, collectively known as $(R_1 + R_2)$. This can be measured with a low resistance reading ohmmeter as described in item 1 of {8.4.1}, and will be necessarily at the ambient temperature which applies at the time. Multiplication by the correction factors given in {Table 8.6} will adjust the resistance to its value at 20°C.

Table 8.6 Correction factors for ambient temperature

Test ambient temperature °C	Correction factor
5	0.94
10	0.96
15	0.98
20	1.00
25	1.02

These will be the normal resistance values and must be adjusted to take account of the increase in resistance of the conductor material due to its increased temperature under fault conditions. This second correction factor depends on the ability of the insulation to allow the transmission of heat, and values will be found in {Table 8.7} for three of the more common types of insulating material.

An alternative method for measuring $(R_1 + R_2)$ is to carry out a loop impedance test at the extremity of the final circuit and to deduct the external loop impedance for the installation (Z_E). Strictly it is not correct to add and subtract impedance and resistance values, but any error should be minimal.

Table 8.7 Temperature correction factors for insulation

Insulation type	Correction factor	
	bunched or as a cable core	not bunched or sheathed cable
p.v.c.	1.38	1.30
85°C rubber	1.53	1.42
90°C thermosetting	1.60	1.48

8.5 Insulation tests

8.5.1 Testing insulation resistance ~ [713-04 and 713-05]

A low resistance between phase and neutral conductors, or from live conductors to earth, will result in a leakage current. This current could cause deterioration of the insulation, as well as involving a waste of energy which would increase the running costs of the installation. Thus, the resistance between poles or to earth must never be less than half of one megohm (0.5 MΩ) for the usual supply voltages. In addition to the leakage current due to insulation resistance, there is a further current leakage in the reactance of the insulation, because it acts as the dielectric of a capacitor. This current dissipates no energy and is not harmful, but we wish to measure the *resistance* of the insulation, so a direct voltage is used to prevent reactance from being included in the measurement. Insulation will sometimes have high resistance when low potential differences apply across it, but will break down and offer low resistance when a higher voltage is applied. For this reason, the high levels of test voltage shown in {Table 8.8} are necessary. {8.7.1} gives test instrument requirements.

Before commencing the test it is important that:

1   electronic equipment which could be damaged by the application of the high test voltage should be disconnected. Included in this category are electronic fluorescent starter switches, touch switches, dimmer switches, power controllers, delay timers, switches associated with passive infra-red detectors (PIRs), RCDs with electronic operation *etc.* An alternative to disconnection is to ensure that phase and neutral are connected together before an insulation test is made between them and earth.

2   capacitors and indicator or pilot lamps must be disconnected or an inaccurate test reading will result.

Table 8.8 Required test voltages and minimum insulation resistance values.

(from [Table 71A] of BS 7671: 1992)

Nominal circuit voltage	Test voltage	Minimum insulation resistance
	(V)	(MΩ)
Extra-low voltage circuits supplied from a safety isolating transformer	250	0.25
Up to 500 V except for above	500	0.5
Above 500 V up to 1000 V	1000	1.0

The insulation resistance tester must be capable of maintaining the required voltage when providing a steady state current of 1 mA.

Where any equipment is disconnected for testing purposes, it must be subjected to its own insulation test, using a voltage which is not likely to result in damage. The result must conform with that specified in the British Standard concerned, or be at least 0.5 MΩ if there is no Standard.

The test to earth {Fig 8.10} must be carried out on the complete installation with the main switch off, with lamps and other equipment disconnected, but with fuses in, circuit breakers closed and all circuit switches closed. Where two-way switching is wired, only one of the two strapper wires will be tested. To test the other, both two-way switches should be operated and the system retested. If desired, the installation can be tested as a whole, when a value of at least 0.5 MΩ should be achieved, *see* {Fig 8.10}. In the case of a very large installation where there are many earth paths in parallel, the reading would be expected to be lower. If this happens, the installation should be subdivided and retested, when each part must meet the minimum requirement.

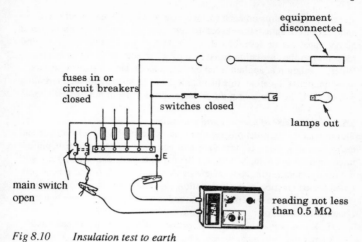

Fig 8.10   Insulation test to earth

Insulation resistance tests between poles must be carried out as indicated in {Fig 8.11} with a minimum acceptable value for each test of 0.5 MΩ. However, where a reading of less than 2 MΩ is recorded for an individual circuit, (the minimum value required by the Health and Safety Executive), there is the possibility of defective insulation, and remedial work may be necessary. As indicated above, tests on SELV and PELV circuits are carried out at 250 V. However tests between these circuits and the live conductors of other circuits must be made at 500 V. Tests to earth for PELV circuits are at 250 V, whilst FELV circuits are tested as LV circuits at 500 V.

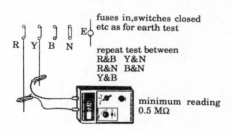

Fig 8.11   Insulation tests between poles

8.5.2   *Tests of non-conducting floors and walls*   ~   [413-04, 471-10 and 713-08]

Where protection against indirect contact is provided by a non-conducting location, the following requirements apply:
1   there must be no protective conductors
2   if socket outlets are used they must not have an earthingcontact
3   it should be impossible for any person to touch two exposed conductive parts at the same time
4   floors and walls must be insulating.

To test this last item and so to make sure that the floors and walls are non-conducting, their insulation has to be tested.

The requirements are shown in {Fig 8.12}, the electrodes used for making contact with floors and walls being a special type which are pressed onto the surface with a force of not less than 750 N (77 kg or 169 lb) for floors or 250 N (26 kg or 56 lb) for walls. The resulting insulation resistance of not less than three points on each surface, one of which must be between 1 m and 1.2 m from an extraneous conductive part (if there is one), measured at 500 V, must not be less than 0.5 MΩ. Attention is drawn to the natural reduction in the insulation resistance of a surface as humidity increases. Where insulation is applied to an extraneous conductive part to provide a non-conducting location, this insulation must be tested with an alternating p.d. of 2 kV. In normal use, the leakage current should not exceed 1 mA.

8.5.3   *Tests of barriers and enclosures*   ~ [713-07]

Throughout the Regulations reference is made to the use of barriers and enclosures to prevent contact with live parts (direct contact). If manufactured equipments comply with the British Standards concerned, they will not need further testing, but where barriers and enclosures have been provided during erection of the installation, they must be tested. Full details of the IP classification system will be found in {Table 2.1}, but the two most common tests are for:

1   IP2X - no contact can be made with a probe 12 mm in diameter and 80 mm long - in other words, a human finger

2   IP4X - no contact can be made with a rod of diameter 1 mm.

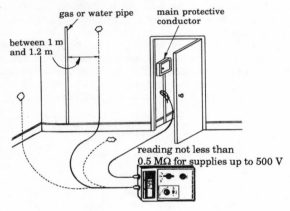

Fig 8.12   Insulation test of floors and walls for non-conducting location

8.5.4   *Tests for electrical separation of circuits*   ~ [413-06 and 713-06]

This section is concerned with tests necessary to ensure the safety of separated extra-low voltage (SELV), protective extra-low voltage (PELV) and functional extra-low voltage (FELV) circuits which are explained in {7.16}. In general, the requirement is a thorough inspection to make sure that the source of low voltage (most usually a safety isolating transformer) complies in all respects with the British Standard concerned, followed by an insulation test between the extra-low voltage and low voltage systems. The test is unusual in that a 500 V dc supply (from an insulation resistance tester) must be applied between the systems for one minute, after which the insulation resistance must not be less than 5 MΩ for SELV or PELV systems, or 0.5MΩ for FELV systems. A further test at 3750 V dc for one minute is passed if no flashover occurs. This test in particular can be dangerous, and special care should be taken.

For SELV and FELV circuits, additional inspection must ensure that the low voltage requirements (not exceeding 50 V ac or 120 V dc) are met. If the voltage exceeds 25 V ac or 70 V dc (60 V ripple-free), barriers and enclosures must be tested to IP2X (*see* {Table 2.1})and a 500 V dc insulation test applied for one minute between the live conductors and metal foil wrapped round the insulation should produce a result of at least 0.5 MΩ

When insulation testing on electrically isolated circuits or on equipment which might be damaged by the test voltage, phase and neutral must be connected together and the test applied between them and earth.

8.6 Earth testing

8.6.1 *Testing earth electrodes* ~ [713-11]

The earth electrode, where used, is the means of making contact with the general mass of earth. Thus it must be tested to ensure that good contact is made. A major consideration here is to ensure that the electrode resistance is not so high that the voltage from earthed metalwork to earth exceeds 50 V. Where an RCD is used, this means that the result of multiplying the RCD operating current (in amperes) by the electrode resistance (in ohms) does not exceed 50 (volts). If a 30 mA RCD is used, this allows a maximum electrode resistance of 1,666 Ω, although it is recommended that earth electrode resistance should never be greater than 220 Ω.

There are several methods for measurement of the earth electrode resistance. In all cases, the electrode must be disconnected from the earthing system of the installation before the tests commence.

1   *Using a dedicated earth resistance tester*

The instrument is connected as shown in {Fig 8.13} with terminals C1 and P1 being connected to the electrode under test (X). To ensure that the resistance of the test leads does not affect the result, separate leads should be used for these connections. If the test lead resistance is negligible, terminals C1 and P1 may be bridged at the instrument and connected to the earth electrode with a single lead.

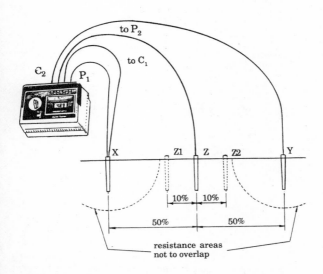

Fig 8.13   *Measurement of earth electrode resistance with a dedicated tester*

Terminals C2 and P2 are connected to temporary spikes which are driven into the ground, making a straight line with the electrode under test. It is important that the test spikes are far enough from each other and from the electrode under test. If their resistance areas overlap, the readings will differ for the reason indicated in {Fig 8.14}. Usually the distance from X to Y will be about 25 m, but this depends on the resistivity of the ground. To ensure that resistance areas do not overlap, second and third tests are made with the electrode Z 10% of the X to Y distance nearer to, and then 10% further from, X. If the three readings are substantially in agreement, this is the resistance of the electrode under test. If not, test electrodes Y and Z must be moved further from X and the tests repeated.

The tester provides an alternating output to prevent electro-lytic effects. If the resistance to earth of the temporary spikes Y and Z is too high, a reduction is likely if they are driven deeper or if they are watered.

2   *Using a transformer, ammeter and voltmeter.*

The system is connected as shown in {Fig 8.15}. Current, which can be adjusted by variation of the resistor R, is passed through the electrode under test (X) to the general mass of earth and hence to the test electrode Y. The voltmeter connected from X to Z measures the volt drop from X to the general mass of earth. The electrode resistance (Ω) is calculated from:

$$\frac{\text{voltmeter reading (V)}}{\text{ammeter reading (A)}}$$

As in the case of the dedicated tester, the test electrode Z must again be moved and extra readings taken to ensure that resistance areas do not over-lap. It is important that the voltmeter used has high resistance (at least 200 Ω/V) or its low resistance in parallel with that of the electrode under test will give a false result.

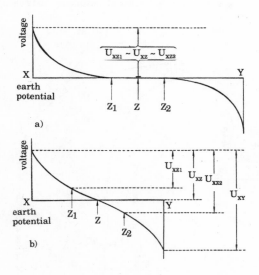

Fig 8.14   *Effect of overlapping resistance areas a) resistance areas not overlapping   b) resistance areas overlapping*

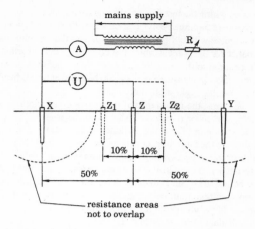

Fig 8.15   Measurement of earth electrode resistance with a transformer, ammeter and voltmeter

3   *Using an earth-fault loop impedance tester*

The tester is connected between the phase at the origin of the installation and the earth electrode under test as shown in {Fig 8.16}. The test is then carried out, the result being taken as the electrode resistance although the resistance of the protective system from the origin of the installation to the furthest point of the installation must be added to it before its use to verify that the 50 V level is not exceeded. If an RCD with a low operating current is used, the protective system resistance is likely to be negligible by comparison with the permissible electrode resistance.

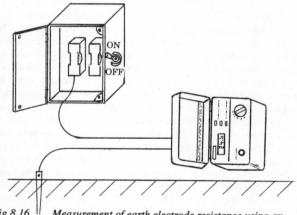

Fig 8.16   Measurement of earth electrode resistance using an earth-fault loop tester

It is most important to ensure that earthing leads and equipotential bonds are reconnected to the earth electrode when testing is complete.

8.6.2   *Measuring earth-fault loop impedance and prospective short-circuit current* ~ [713-10]

The nature of the earth-fault loop and its significance have been considered in detail in {5.3}. Since the loop includes the resistance of phase and protec-

181

tive conductors within the installation, the highest values will occur at points furthest from the incoming supply position where these conductors are longest. A measurement within the installation will give the complete earth-fault loop impedance for the point at which it is taken (Z_S), or the earth-fault loop impedance external to the installation (Z_E) may be measured at the supply position. Internal loop measurements should be taken at points furthest from the intake to give the highest possible results.

In simple terms, the impedance of the phase-to-earth loop is measured by connecting a resistor (typically 10 Ω) from the phase to the protective conductor as shown in {Fig 8.17}. A fault current, usually something over 20 A, circulates in the fault loop, and the impedance of the loop is calculated within the instrument by dividing supply voltage by the value of this current. The resistance of the added resistor must be subtracted from this calculated value before the result is displayed. An alternative method is to measure the supply voltage both before and whilst the loop current is flowing. The difference is the volt drop in the loop due to the current, and loop impedance is calculated from voltage difference divided by current.

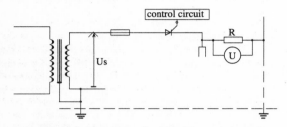

Fig 8.17   Simple principle of earth-fault loop testing

Since the loop current is very high, its duration must be short and is usually limited to two cycles (or four half-cycles) or 40 ms for a 50 Hz supply. The current is usually switched by a thyristor or a triac, the firing time being controlled by an electronic timing circuit. It is very important to have already checked the continuity of the protective system before carrying out this test. A break in the protective system, or a high resistance within it, could otherwise result in the whole of the protective system being directly connected to the phase conductor for the duration of the test. Commercial testers are usually fitted with indicator lamps to confirm correct connection or to warn of reversed polarity. {Fig 8.18} shows a typical earth-fault loop tester connected to a socket outlet so that its loop impedance can be measured. If the circuit to be measured includes socket outlets, the tester is connected as indicated in {Fig 8.18}. Special leads for connection to phase and to earth are provided by suppliers for all other circuits.

Tests must be carried out at the origin of the installation, at each distribution board, at all fixed equipment, at all socket outlets, at 10% of all lighting outlets (choosing points furthest from the supply) and at the furthest point of every radial circuit.

A modified version of the earth-fault loop tester, which effectively measures the phase to neutral impedance and calculates then displays the value of the current which would flow if the supply voltage were applied to this impedance are readily available. The principle of such a PSC tester is described in {3.7.2}.

Since the test result is dependent on the supply voltage, small variations will affect the reading. Thus, the test should be repeated several times to ensure consistent results. The test resistor will be connected across the mains for the duration of each test, and will become very hot if frequent tests are made. Some testers will then 'lock out' to prevent further testing until the resistor temperature falls to a safe value.

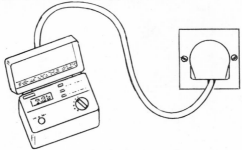

Fig 8.18   Earth-fault loop tester connected for use

The earth fault loop impedance measured as described will be for installation cables at normal operating temperature. Under fault conditions, cable temperature will rise, and so will the resistive component of the impedance. This effect can be taken into account by applying the ambient temperature and insulation material correction factors shown in {Tables 8.6 and 8.7}. An alternative is to ensure that the measured values of earth fault loop impedance do not exceed three quarters of the maximum values shown in {Tables 5.1, 5.2 or 5.4} as appropriate.

The effect of supply voltage on the calculation of earth fault loop impedance is considered in {5.3.4}.

A circuit protected by an RCD will need special attention, because the earth-fault loop test will draw current from the phase which returns through the protective system. This will cause an RCD to trip. Therefore, any RCDs must be by-passed by short circuiting connections before earth-fault loop tests are carried out. It is, of course, of the greatest importance to ensure that such connections are removed after testing. One manufacturer supplies a patented loop tester which does not require RCDs to be short circuited and which will not cause them to trip when the earth-fault loop test is made.

When loop testing at lighting units controlled by passive infra-red detectors (PIRs), there may be damage to the associated electronic switches unless they are short-circuited before testing.

8.6.3   *Testing residual current devices (RCDs)*   ~   [713-12]
Residual current devices should comply with BS 4293 and are described in {5.9}, from which it will be seen that they are provided with a built-in self test system which is intended to be operated regularly by the user. The Regulations require that correct operation of this test facility should be checked, and that other tests are also carried out. The time taken for the device to operate must be measured, so the old type of 'go, no-go' tester is no longer adequate. {8.7.1} gives test instrument requirements.

RCD tests are carried out with a special tester which is connected between phase and protective conductors on the load side of the RCD after disconnecting the load {Fig 8.19}. A precisely measured current for a carefully timed period is drawn from the phase and returns *via* the earth, thus tripping the device. The tester measures and displays the exact time taken for the circuit to be opened. This time is very short, in most cases being between 10 and 20 ms, although it can be much longer, especially for S-types which have delayed operation.

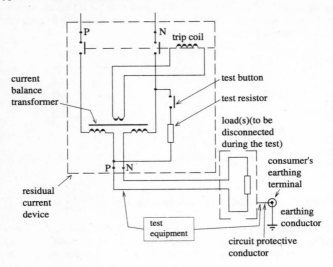

Fig 8.19   Connections for an RCD tester

1   *General purpose non-delayed RCDs*

This is a general purpose type of RCD which is intended to operate very quickly at its rated current. Three tests are required:

a) 50% of the rated tripping current applied for 2 s should *not* trip the device,

b) 100% of rated tripping current, which should not be applied for more than 2 s, must cause the device to trip within 200 ms (0.2 s), and

c) where the device is intended to provide supplementary protection against direct contact, a test current of 150 mA, applied for no more than 50 ms, should cause the device to operate within 40 ms.

2   *Time-delayed RCDs*

In {5.9.2} we discussed the need for discrimination between RCDs. This type is deliberately delayed in its operation to make sure that other devices which are connected downstream of it will operate more quickly. A 3:1 discrimination ratio is required between two RCDs which are connected in series, and this must be verified before testing. It means that the delayed RCD must have an operating current at least three times that of the non-delayed type. For example, to discriminate properly with a 30 mA device, a second connected on the supply side would need to have an operating current of at least 90 mA (in practice, a 100 mA RCD is likely to be used).

The test for the time-delayed RCD consists of applying 100% of the normal rated current, when the device should trip within the time range of:

50% of rated time delay plus 200 ms, and

100% of rated time delay plus 200 ms.

For example, an RCD with a rated tripping time of 300 ms should trip within a time range of:

(150 + 200) ms = 350 ms

and   (300 + 200) ms = 500 ms.

An RCD tester is an electronic device which draws current from the supply for its operation. This current is usually of the order of a few milliamperes which is taken from the phase and neutral of the supply under test, and will have no effect on the measurement of single-phase systems. However, if a three-wire three-phase system (there is no neutral with this supply) is being tested, the tester must be connected to a neutral conductor to provide the power it needs for operation. Thus, its operating current will flow through a line conductor and return through the neutral, giving a basic imbalance. A 'no-trip' test must also be carried out, during which the RCD must not operate when 50% of the rated tripping current is applied for 2 s. The extra current to power the tester, which adds to the test current, may then cause operation. It is necessary in this case to obtain from the RCD manufacturer the value of this current and to take it into account before failing a device on the 50% test.

The RCD tester is connected to the device to be tested by plugging it into a suitable socket outlet (*see* {Fig 8.20}) or by connecting to phase and neutral with special leads obtainable from the instrument supplier.

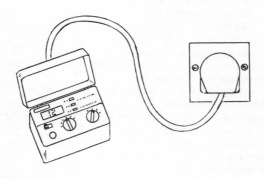

Fig 8.20   *RCD tester connected for use*

8.7   Test instrument requirements
8.7.1   *Basic requirements*

Guidance Note 3 - Inspection and Testing - makes it clear that instrument accuracy is required to be at least that shown later in this sub-section for the various types of instrument. Regular recalibration using standards traceable to National Standards is now required, together with checking after any incident which has involved mechanical mishandling. Many electrical installers will not be used to sending their instruments regularly for recalibration, but *must* now do so. Guidance Note 3 is not specific on the time intervals at which recalibration must be carried out, but it would seem sensible for occasionally used instruments to receive attention every two years, whilst those used frequently are likely to need recalibration annually.

If installations are to be tested to show that they comply with the 16th Edition, the following instruments will be necessary. After the name of the instrument are brief notes which may be helpful in choosing a new instrument or in deciding if one already to hand will be satisfactory. The first four instruments listed are absolutely essential for all tests, although the low resistance tester and the insulation resistance tester may be combined in a single instrument. The last two instruments will not often be required on simple installations, since applied voltage and earth electrode resistance tests are often not needed. All instruments used should conform to the appropriate British Standard safety specification (BS 4743 for electronic types, and BS 5458 for electrical instruments).

Low resistance ohmmeter
        Basic instrument accuracy required is ±2%
        Test voltage ac or dc, between 3 V and 24 V
        Test current not less than 20 mA
        Able to measure to within 0.01 Ω (resolution of 0.01)
        May be the continuity range of an insulation resistance tester.

Insulation resistance tester
        Direct test voltage depends on the circuit under test, but will be:-
                250 V for extra-low voltage circuits
                500 V for other circuits supplied at up to 500 V
                1000 V for circuits rated between 500 V and 1000V.
        Must be capable of delivering a current of 1 mA at the minimum allowable resistance level, which is:-
                250 kΩ for the 250 V tester
                500 kΩ for the 500 V tester
                1 MΩ for the 1,000 V tester
        Basic instrument accuracy required is ±2%
        Must have a facility to discharge capacitance up to 5 µF which has become charged during the test
        May be combined with the low resistance ohmmeter

Earth-fault loop impedance tester
        Must provide 20 to 25 A for up to two cycles or four half-cycles
        Basic instrument accuracy required is ±2%
        Able to measure to within 0.01 Ω (resolution of 0.01)

Residual current device (RCD) tester
        Must perform the required range of tests (*see* {8.6.3})
        Suitable for standard RCD ratings of 6, 10, 30, 100, 300 and 500 mA
        Must NOT apply full rated test current for more than 2 s
        Currents applied must be accurate to within ±10%
        Able to measure time to within 1 ms (resolution of 1)
        Must measure opening time with an accuracy of ±5%

Applied voltage tests
        Must apply a steadily increasing voltage measured with an accuracy of ±5%
        Must have means of indicating when breakdown has occurred
        Must be able to maintain the test voltage for at least one minute
        Maximum output current must not exceed 5 mA
        Maximum output voltage required is 4,000 V

Earth electrode resistance tester
        Basic instrument accuracy required is ±2%
        Must include facility to check that the resistance to earth of tempo rary test spikes are within limits
        Able to measure to within 0.01 Ω (resolution of 0.01)

8.7.2   *Accuracy and resolution*
The sub-section above has indicated the levels of accuracy and resolution required of the instruments needed to test an electrical installation. The purpose of this sub-section will be to explain the meaning of these two terms.

Accuracy

This term describes how closely the instrument is able to produce and display a correct result, and is usually expressed as a percentage. For example, if a voltage has a true level of 100 V and is measured by a voltmeter as 97 V, this is an error of -3 V. Expressed in terms of the true voltage, it is an error of three volts in one hundred volts, or three per cent (-3%). In this case the reading is low, so the true error would be -3%. Had the error been +3% the reading would have been 103 V. In most cases we do not know if the reading is high or low, so the error is expressed as a percentage which may be positive or negative. Thus, if the voltmeter gave a reading of 100 V but was known to have an accuracy of ±4%, the actual voltage could lie anywhere in the range from:

$$100 + (4/100) \times 100 \text{ V} \quad \text{to} \quad 100 - (4/100) \times 100 \text{ V}$$
or $\quad 100 + 4$ V $\quad\quad\quad$ to $\quad 100 - 4$ V

which is between 104 V and 96 V.

The values given in the Guide are called *basic instrument accuracies* which indicate the possible error with the instrument itself. In practice, there are many factors which affect the value which is to be measured, and which will further reduce the accuracy. These are divided into two types.

Instrument errors are largely due to the fact that the true error is not constant, varying from point to point over the instrument range. Other factors, such as battery voltage, ambient temperature, operator's competence, and the position in which the instrument is held or placed (such as vertical or horizontal) will also affect the reading.

Field errors concern external influences which may also reduce accuracy, and may include capacitance in the test object, external magnetic fields due to cables and equipment, variations in mains voltage during the test period, test lead resistance, contact resistance, mains pickup, thermocouple effects, and so on.

It is important to appreciate that percentage accuracy is taken in terms of the full scale reading of an analogue instrument, or the highest possible reading of a digital type. Thus, in a multirange instrument, it is related to the scale employed, not to the reading taken. For example, if an ohmmeter with a known error of ±5% is on its 100 Ω scale and reads 8 Ω, the true reading will lie between

$$8 + \frac{100 \times 5}{100} \ \Omega = 8 + 5 \ \Omega = 13 \ \Omega \text{ and}$$

$$8 - \frac{100 \times 5}{100} \ \Omega = 8 - 5 \ \Omega = 3 \ \Omega$$

and *not* between $8 \pm \dfrac{8 \times 5}{100} \ \Omega = 8 \pm 0.4 \ \Omega$ or 7.6 Ω and 8.4 Ω

Thus it can be seen that the highest accuracy will result from using the lowest possible scale on a multirange instrument.

Resolution

This term deals with the ability of an instrument to display a reading to the required degree of accuracy. For example, if we were measuring the earth-fault loop impedance of a socket outlet circuit protected by a 30 A miniature circuit breaker type 2, we would need to ensure that the impedance value was not more than 1.14 Ω {Table 5.2}. If this were done using a digital meter with three digits and a lowest range of 99.9 Ω, we could obtain a reading of 1.1 Ω or another of 1.2 Ω, but not 1.14 Ω. This would indicate that the instrument resolution was to the nearest 0.1 Ω, which usually is not close enough for electrical installation measurements. If the same three-digit instrument had a lower scale of 9.99 Ω, it would be capable of reading 1.14 Ω and would have a resolution of 0.01.

8.8 Supporting paperwork

8.8.1 Why bother with paperwork? ~ [721-1, 743-1, 741]

When an installation is complete, including additions or alterations to an existing installation, the persons responsible for the work must report to the owner that it is complete and ready for service. This is in the form of a completion and inspection certificate which must be separately signed to verify the design, the construction and the inspection and test aspects (*see* {8.8.2}) to confirm that the installation complies with the Regulations.

This certificate will verify that the installation is safe to use and ready for service, and should be signed by a competent person who should preferably be one of the following:

1	a professionally qualified electrical engineer, or
2	a member of the ECA (Electrical Contractors' Association), or
3	a member of the ECA of Scotland, or
4	an approved contractor of the NICEIC (National Inspection Council for Electrical InstallationContracting), or
5	a qualified person acting on behalf of one of the above.

In all cases, the Certificates must state for whom the qualified person is acting.

The installer should also compile an operational manual for the installation, which will include all the relevant data, including:

1	a full set of circuit and schematic drawings,
2	all design calculations for cable sizes, cable volt drop, earth-loop impedance, *etc.*
3	leaflets or manufacturers' details for all the equipment installed,
4	'as fitted' drawings of the completed work where applicable,
5	a full specification,
6	copies of the completion and inspection certificate, together with any other commissioning records,
7	a schedule of dates for periodic inspection and testing,
8	the names, addresses and telephone numbers of the designer, the installer, and the inspector/tester.

This requirement will be new to many electrical installation installers. Its rationale is to ensure that future owners or users of the installation, as well as those who maintain it or who may modify it, have full information. If difficulty is experienced in preparing the operational manual, reference to BS 4884 and to BS 4940, may be helpful.

8.8.2 Completion and inspection certificate ~ [742]

Following completion of inspection and testing, a completion and inspection certificate, together with particulars of the electrical installation and a schedule of test results, must be provided to the person who ordered the work. It must be signed three times (by the designer, the installer and the inspector/tester) to certify that the installation has been designed, constructed, inspected and tested in accordance with the 16th Edition of the IEE Wiring Regulations. The form of the certificate giving particulars of the electrical installation and of completion and inspection is given as {Table 8.9}. In some cases, where there are good reasons and where a qualified electrical engineer has given his approval, the installation may not comply fully with the Regulations. In such a case, full details of the departures must be stated on the completion and inspection certificate as indicated in {Table 8.9}, although this table may be expanded and redesigned as necessary to cover the particular installation under test. The recommended intervals between periodic inspections and tests should be as indicated in {Table 8.4}.

8.8.3   *Installation alterations and additions* ~ [721, 743]

The changes in the occupation and uses of situations where there are electrical installations make alterations and additions a common occurrence. Before commencing alterations and additions to an installation it is very important to verify that the existing supply system, installation and earthing arrangements are capable of feeding the proposed new installation safely. For example, the additional installation in a large extension to a house may impose extra loads on the supply system which it is incapable of meeting. It is the responsibility of the person carrying out the extra work to ensure that it, as well as the existing installation, will function safely and correctly. The tester must check that all materials made redundant by the installation changes, such as cables, wooden pattresses *etc.*, and which may be responsible for the spread of fire, are removed. Additionally, the inspector must look for other prospective fire risks, such as:

1.   flexible cords not securely held by cord grips,
2.   a dangerous increase in the use of adaptors,
3.   worn or otherwise damaged flexible cords,
4.   signs of overheating of appliances, plugs and connectors,
5.   over-rated lamps fitted to luminaires, and
6.   heaters with insecure or missing guards.

It is *not* his responsibility to rectify defects and faults in the existing installation, but he *must* test and inspect it, reporting deviations on the completion and inspection certificate which is provided on completion of the work. Should dangerous defects be found in an existing installation it would be clearly irresponsible of the person carrying out the alterations or extensions not to bring them to the urgent attention of the user.

8.8.4   *Periodic inspection and testing* ~ [514-12-02, 731 and 744]

The importance of regular inspection and testing of electrical installations cannot be overstated, but unfortunately it is an aspect of electrical safety which is very often overlooked. It is now a requirement of the Regulations that the installer of an installation must tell the user of the need for periodic test and inspection and the date on which such attention is necessary. Probably the good contractor will institute a system so that a reminder is sent to the customer at the right time. Suggested intervals between inspections and tests are shown in {Table 8.4}. The results of sample tests should be compared with those taken when the installation was last tested and any differences noted. Unless the reasons for such differences can be clearly identified as relating only to the sample concerned, more tests must be carried out. If these, too, fail to comply with the required values, the complete installation must be retested and the necessary correcting action taken.

The inspection and testing must be carried out with the same degree of care as is required for a new installation. In fact, more care is often needed because dangers can occur to the testers and to others in the situation in the event of failure of parts of an installation such as the protective system. The tester must look out for additions to the installation, or for changes in the use of the area it serves, either of which may give rise to fire risks. Included may be the addition of thermal insulation, the installation of additional cables in conduit or trunking, dust or dirt which restricts ventilation openings or forms an explosive mixture with air, changing lamps for others of higher rating, missing covers on joint boxes and other enclosures so that vermin may attack cables, and so on.

Table 8.9   ELECTRICAL INSTALLATION COMPLETION CERTIFICATE   (BS 7671: 1992)

DETAILS OF THE INSTALLATION

Client's name/title ....

Installation address ....

New installation   □

Extent of installation covered by this certificate

Addition to existing installation   □

Alteration to existing installation   □
(Use continuation sheet if necessary)

PARTICULARS OF THE INSTALLATION

Type of earthing   TN-C-S □ TN-S □   TT □ TN-C □   IT □

Details of earth electrode: Type ....

Location ....

Method of measurement ....

Resistance ....(Ω)

Characteristics of the supply at the origin of the installation

Nominal voltage .... Frequency .... Hz

No. of phases ....

Maximum demand (load) .... A per phase

□   Measured   □   Calculated   □   Other

Maximum prospective fault current (kA)..................................

External earth fault loop impedance (Ω)................................

Overcurrent protective device at origin: Type BS.............Rating ....A

Main switch or circuit breaker:

Number of poles .... Type BS .... Rating ....A

(if a residual current device, rated residual operating current .... mA)

Method of protection against indirect contact:

1. Earthed equipotential bonding and automatic disconnection of supply

2. Other (describe) ....

Main equipotential bonding conductors: material .... csa ....mm^2

Schedule of Circuits and Test Results : see pages .... to ....

Comments on existing installation, in case of an alteration or addition

DESIGN

I/we being the person(s) responsible (as indicated by my/our signatures below) for the design of the electrical installation, particulars of which are described on page 1, CERTIFY that the said work for which I/we have been responsible is to the best of my/our knowledge and belief in accordance with BS 7671: 1992 - Requirements for Electrical Installations (16th Edition IEE Wiring Regulations) amended to .... except for the departures, if any, stated in this certificate.

Details of departures (if any) from BS 7671 1992 (120-02)

The extent of liability of the signatory is limited to the work described on page 1 of this form as the subject of this Certificate.

For the DESIGN of the installation
Name (in BLOCK letters)...................................Position ....
Signature ....
Date ....
For and on behalf of ....
Address
....

CONSTRUCTION

I/we being the person(s) responsible (as indicated by my/our signatures below) for the construction of the electrical installation, particulars of which are described on page 1, CERTIFY that the said work for which I/we have been responsible is to the best of my/our knowledge and belief in accordance with BS 7671: 1992 - Requirements for Electrical Installations (16th Edition IEE Wiring Regulations) amended to .... except for the departures, if any, stated in this certificate.

The extent of liability (if any) from BS 7671 1992 (120-02) specified by the designer but not indicated in the design certificate above

For the CONSTRUCTION of the installation
Name (in BLOCK letters) ....Position ....
Signature ....Date ....
For and on behalf of ....
Address ....

INSPECTION AND TEST

I/we being the person(s) responsible (as indicated by my/our signatures below) for the inspection and test of the electrical installation, particulars of which are described on page 1, CERTIFY that the said work for which I/we have been responsible is to the best of my/our knowledge and belief in accordance with BS 7671: 1992 - Requirements for Electrical Installations (16th Edition IEE Wiring Regulations) amended to .... except for the departures, if any, stated in this certificate.

The extent of liability of the signatory is limited to the work described as the subject of this Certificate.

For the INSPECTION AND TEST of the installation
Name (in BLOCK letters) .... Position ....
Signature ....Date ....
For and on behalf of ....
Address ....

I/we RECOMMEND that this installation be further inspected and tested after an interval of not more than .... months/years

The sequence of periodic tests differs slightly from that for initial tests {8.3.2} because in this case the supply will always be connected before testing starts. The tests required and the sequence in which they must be performed are shown in {Table 8.13}. There will be cases where necessary information in the form of charts, diagrams, tables, *etc.*, is not available. The tester will then need to investigate the installation more thoroughly to make sure that he is fully conversant with it before he can carry out his work.

It must be clearly understood that a retest of a working installation may not be as full as that carried out before the installation was put into service. For example, it may be impossible to switch off certain systems such as computers, and if full testing is not possible, this should be made clear on the certificate. In such cases, the installation should be carefully inspected to ensure that possible dangers become apparent. In some cases it may be necessary to carry out measurements, such as the value of the earth leakage current, which will indicate the health of the system without the need to disconnect it.

A sample of 10% of switching devices must be thoroughly internally inspected and tested. If results are poor, the procedure must be extended to include all switches. The condition of conductor insulation and other protection against direct contact must also be inspected at all distribution boards and at samples of switchgear, luminaires, socket outlets, *etc*. There should be no signs of damage, overloading or overheating.

Protective and equipotential bonding conductors must not be disconnected from the main earthing terminal unless it is possible first to isolate the supply. The tester must ensure that a durable notice to the wording given in {Table 8.12} is fixed at the mains position.

On completion a full inspection and testing. Additional matters to report are: Additional matters to report are:-

1   full test results to enable comparison with earlier tests, from which the rate of deterioration of the installation (if any) can be assessed. A suggested form, is given as {Table 8.10}
2   the full extent of the parts of the system tested — notes of omissions may be very important
3   any restrictions which may have been imposed on the tester and which may have limited his ability to report fully
4   any dangerous conditions found during testing and inspection, non-compliance with the Regulations, or any variations which are likely to arise in the future.

Table 8.10 PERIODIC INSPECTION REPORT FOR AN ELECTRICAL INSTALLATION (BS 7671: 1992)

DETAILS OF THE CLIENT

Client ....

Address ....

Purpose for which the report is required

....

....

(continued)

DETAILS OF THE INSTALLATION

Occupier ....Address ....
....
Description of premises: ☐ Domestic ☐ Commercial ☐ Industrial
Other ....
Estimated age of the electrical installation .... years
Evidence of alterations or additions ☐ Yes ☐ No ☐ Not apparent
If "yes", estimated age .... years.
Date of last inspection ....
Records available ☐ Yes ☐ No
Records held by ....
Type of earthing TN-C-S ☐ TN-S ☐ TT ☐ TN-C ☐ IT ☐
Details of earth electrode:
Type .... Location ....
Method of measurement ....
Resistance ....(Ω)
Characteristics of the supply at the origin of the installation
Nominal voltage .... Frequency .... Hz No. of phases ....
Maximum demand (load) .... A per phase
☐ Measured ☐ Calculated ☐ Other
Maximum prospective fault current (kA)
External earth fault loop impedance (Ω)
Overcurrent protective device at origin:
Type BS............. Rating .... A
Main switch or circuit breaker:
Number of poles .... Type BS .... rating ....A
(if a residual current device, rated residual operating current .... mA)
Method of protection against indirect contact:
1. Earthed equipotential bonding and automatic disconnection of supply☐
2. Other (describe) .... ☐

Main equipotential bonding conductors:
material .... csa ....mm^2

EXTENT AND LIMITATIONS OF THE INSPECTION

Extent of electrical installation covered by this report
....
Limitations ....

RECOMMENDATIONS

Referring to the "Schedule(s) of Inspection and Test Results", and subject to the limitations specified,

No remedial work is required ☐

or the following items need ☐

....

....

....

One of the following numbers shall be placed alongside each of the items detailed above:-

1. require urgent attention 2. require improvements 3. requires further investigation 4. does not comply with the current BS 7671(as amended). (This does not necessarily imply that the electrical installation is unsafe).

SUMMARY OF THE INSPECTION

Date(s) of inspection ....

General condition of the installation ....

....

Overall assessment: Satisfactory/Unsatisfactory

Schedule of the inspection: See sheets..........attached

Schedule of tests: See sheets ....attached

Schedule of inspection and test results: See sheets .... attached

NEXT INSPECTION

We recommend that the installation should be re-inspected after an interval of not more than .... months/years

DECLARATION

To the best of our knowledge and belief I/we confirm that the details recorded above and in the attached Schedule(s) of Inspection and Test Results and the Recommendations (F) are an accurate assessment, within the limits specified at D, of the condition of the electrical installation at B.

INSPECTED BY:-   REVIEWED BY:-

Signature ....   Signature ....

Name (Capitals) ....   Name (Capitals) ....

Date ....   Date ....

For and on behalf of ....

Address ....

....

Table 8.11 Schedule of test results

Resistance readings for continuity tests on ring final circuits
.... (list all)
Continuity of protective conductors and equipotential bonding checked
YES/NO
Insulation resistance of installation to earth .... (list all)
Insulation resistance of installation between poles .... (list all)
Insulation resistance to earth of equipment disconnected during main test
.... (list all)
Protection against direct contact by barriers and enclosures checked   YES/NO
Resistance of non-conducting floors and walls .... (list all)
Polarity check on single-pole switches and protective devices   YES/NO
Earth-fault loop impedance tests for all outlets .... (list all)
Operation of residual current devices, operating time ....
(list all)

Condition of flexible cables and cords, switches, plugs and socket outlets checked
YES/NO

Where an installation was constructed to comply with an earlier Edition of
the Regulations, tests should be made as required by the 16th Edition as far
as it is applicable and the position fully explained, with suggestions of necessary action, in the Inspection and Test report.

Table 8.12   Notice - periodic inspection and testing
(from [514-12-01] of BS 7671: 1992)
IMPORTANT
This installation should be periodically inspected and tested and a report on
its condition obtained, as prescribed in BS 7671 (formerly The IEE Wiring
Regulations for Electrical Installations) published by the Institution of Electrical Engineers.

Date of last inspection ....

Recommended date of next inspection ....

Table 8.13 Sequence of periodic testing
(from paragraph 6.2 of IEE Guidance Note 3
Inspection and Testing)

1	Continuity of protective conductors and earthed equipotential bonding.
2	Polarity.
3	Earth fault loop impedance.
4	Insulation resistance.
5	Operation of switches and isolators.
6	Operation of residual current devices.

together with the following where appropriate

7	Continuity of ring final circuit conductors.
8	Earth electrode resistance.
9	Manual operation of circuit breakers.
10	Electrical separation of circuits.
11	Insulation resistance of non-conducting floors and walls.

Pads of Certificates such as these are available from
E·P·A Press, PO Box 41, Saffron Walden, CB11 4LJ
Tel 01799 541207   Fax 01799 541166

Cross reference index

Note that when a two-part Regulation number is used, all Regulations included in that group are included. For example, if reference is made to 553-03, this must be taken to include Regulations 553-03-01 to 553-03-04 inclusive.

16th Edition	Electrician's Guide	16th Edition	Electrician's Guide
PART 1			5.2.5, 5.3.4 & 7.8.2
110-01 to 110-04	2.2.1		
120-01 to 120-03	2.2.2	413-02-12	3.4.6, 5.2.3 to 5.2.5, 5.3.4, 5.3.5, 5.4.5 & 7.8.2
120-04 & 120-05	2.2.3		
130-01	2.2.4		
130-02-01	2.2.4, 2.4.5	413-02-13 & 413-02-14	3.4.6, 5.2.3 to 5.2.5, 5.3.4, 5.3.5, & 7.8.2
130-02-02 to 130-02-05	2.2.4		
130-03	2.2.4, 3.5.2, 3.6.1 & 3.8.1		
		413-02-15	3.4.6, 5.2.3 to 5.2.5, 5.4.3, 5.9.3 & 7.8.2
130-04-01 to 130-04-03	2.2.4 & 5.1.1		
130-04-04	2.2.4, 5.1.1 & 5.4.3	413-02-16 & 413-02-17	3.4.6, 5.9.3 & 7.8.2
130-05-01	2.2.4		
130-05-02	2.2.4 & 3.2.1	413-02-18 to 413-02-20	3.4.6, 5.2.2 & 7.8.2
130-06-01	2.2.4 & 3.2.2		
130-06-02	2.2.4, 3.2.2 and 7.15.1	413-02-21 to 413-02-26	3.4.6, 5.2.6 & 7.8.2
130-07 & 130-08-01	2.2.4	413-02-27 & 413-02-28	3.4.6 & 5.4.3
130-08-02	2.2.4 and 3.5.1	413-03	3.4.6 & 5.8.1
130-09 & 130-10	2.2.4	413-04	3.4.6 & 5.8.2
PART 2		413-05	3.4.6 & 5.8.3
Complete	2.3	413-06	3.4.6, 5.8.4 & 8.5.4
PART 3			
300-01	2.4.1	421-01-01	5.3.1
311-01	2.4.2, 6.2.1 and 6.2.2	422-01	3.5.2 & 6.5.1
		423-01	3.5.3
312-01 and 312-02	2.4.2	424	3.5.2 & 3.5.3
312-03	2.4.2 and 5.2.1	431-01	3.6.1
313-01	2.4.2 and 3.7.2	432-01	3.6.1
313-02	2.4.2	432-02	3.6.2, 3.7.1, 3.7.2 & 3.8.1
314-01	2.4.2 & 6.1		
32	2.4.3	432-03	3.6.2
331-01	2.4.3 & 6.6.1	432-04	3.7.1 & 3.7.2
341-01	2.4.4 & 2.4.5	433-01	3.6.2, 3.6.5, 4.3.8 & 6.2.2
PART 4			
400-01 & 400-02	3.4.2	433-02	3.6.2, 3.6.5, 4.3.8 & 6.2.2
410-01	3.4.4		
411-01	3.4.4 & 7.16.1	433-03	3.6.2, 3.8.4, 4.3.8 & 6.2.2
411-02	3.4.4 & 7.16.2		
411-03	3.4.4 & 7.16.3	434-01	3.7.1, 3.7.3 & 4.3.8
411-04	3.4.4		
412-01 to 412-05-03	3.4.5	434-02	3.7.1, 3.7.2, 3.8.4 & 4.5.8
412-05-04	3.4.5 & 3.4 7		
412-06	3.4.5, 3.4.7 & 5.9.3	434-03-01 & 434-03-02	3.7.1 to 3.7.3 & 4.3.8
		434-03-03	3.7.1, 3.7.3, 4.3.2 & 4.3.8
413-01 to 413-02-03	3.4.6		
413-02-04 & 413-02-05	3.4.6, 8.4.1 & 8.5.2	434-04	3.7.1, 3.8.4 & 4.3.8
413-02-06	3.4.6 & 5.2.3 to 5.2.5	435-01	3.7.5, 3.8.1 & 3.8.9
413-02-07 to 413-02-11	3.4.6, 5.2.3 to	436-01	3.8.2

* There is no 742-01, so it appears that an error has been made in numbering the Regulations. It is expected that at the first opportunity a change will be made so that 742-02 becomes 742-01.

Cross reference Index for the On-Site Guide (OSG), 1st Edition

3.3	3.8	10.3.1	8.4.1	5.2	3.2.4	App 1	6.2
3.4	3.4	10.3.2	8.4.2	5.3	3.2.3	App 2	5.3
3.5	3.7.3	10.3.3	8.4.3	5.4	3.2.2	App 3	4.2
3.6	5.9	10.3.4	8.5	6.1	5.5.2, 5.9.3		
4.1	5.4.3	10.3.5	8.6.1		& 8.2.1	App 4	4.4.1
4.2	5.5.1	10.3.6	8.6.2	7.1	6.3.2	App 5	4.5.3
4.3	5.5.2	11.1	8.6.3	7.2	6.3, 6.5	App 6	8.4.4
4.4	5.2	11.2	8.6.3	7.3	6.6.1, 6.6.2	App 7	4.3
5.1	3.3	11.3	5.9.2	8.1	7.2.1, 7.2.2	App 8	8.8
				8.2	7.14.1, 7.14.2	App 9	Chap 6

2nd Edition of the On-Site Guide

References are the same as shown above for the 1st Editon, except for the following:

4.2	5.4.3	8.4	7.8.1, 7.8.2
4.3	5.4.1	11.3	8.6.3
4.4	5.4.3	11.4	8.6.3
4.5	5.5.1	11.5	8.6.3
4.6	5.5.2	11.6	5.9.2
4.7	5.2	App 6	4.3
8.2	7.2.2	App 7	8.8
8.3	7.14.1, 7.14.2	App 8	Chap 6
App 9	8.4.4	App 10	5.4.5

Cross reference Index for Guidance Note 1 (GN1), 1st Edition
Selection and Erection

G.N.1	Guide	G.N.1	Guide
1.1	2.1	5.7	7.13.1, 7.13.2
1.2	2.2.2, 2.2.3, 2.2.4	6.1	4.3
2.1	Chaps 2, 3 and 4	6.2	6.2.2
2.2	2.4.1	6.3	4.3
2.3	2.5	6.4	4.3.11
2.4	2.4.2, 2.4.3, 2.4.5	7.1	4.4.3, 4.5.1
2.5	6.1	7.2	4.3.6
2.6	2.4.3, 8.8.1	7.3	4.3.6
2.7	2.4.4, 5.10	7.4	5.4.3, 6.6.1, 6.6.2
3.1	3.4	7.5	6.6.3
3.2	3.6	8.1	7.8.2
3.3	3.6.3, 3.6.4	8.2	7.5.2
3.4	3.7.6	8.3	7.11.2
3.5	5.9	8.4	2.2.5
3.6	5.3	8.5	7.15.2
4.1	2.4.3	8.6	2.4.3, 2.4.5
4.2	4.3.4	App A	4.5.3
4.3	4.3.4	App B	2.4.3
4.4	2.4.3	App C	2.4.3
4.5	2.4.3, 4.2.5, 4.5.1	App D	5.2
4.6	2.4.3	App E	6.3, 6.4, 6.5.2
4.7	2.4.3, 4.2.5	App F	5.3, 5.4
4.8	2.4.3, 6.3.1	App G	5.3.6, 8.4.4
5.1	4.2.4, 4.4.1, 7.13.3	App H	4.2.1, 4.2.2
5.2	7.6.2	App I	4.4, 4.5.2, 7.13.2
5.3	4.5.3	App J	6.2.1, 6.2.2
5.4	4.4.1	App K	6.6.1
5.5	4.2.3, 7.5.1, 7.9.2	App L	5.9.3
5.6	4.3.13	App M	-

2nd Edition of Guidance Note 1

G.N.1	Guide		
1.1	2.1	2..5	3.3.5, 4.6.2, 4.6.3, 8.2.1
1.2	2.2.2, 2.2.3, 2.2.4	2.6	2.4.3
1.3	2.2.2, 2.2.4	2.7	2.4.4, 5.10
1.4	2.2.2	2.8	not applicable
2.1	Chaps 2, 3 and 4	3.1	3.4, 3.6
2.2	2.4.1	3.2	3.6
2.3	2.5	3.3	3.6.3, 3.7.2
2.4	2.4.2, 2.4.3, 2.4.5		

Cross reference Index for Guidance Note 2 (GN2), 1st Edition
Isolation and Switching

2nd Edition of Guidance Note 2

Cross reference Index for Guidance Note 3 (GN3), 1st Edition
Inspection and Testing

G.N.3	Guide	G.N.3	Guide
1.2	8.1.2	5.3.10	8.5.1, 8.5.3
1.3	8.1.3	5.3.11	8.1.3, 8.2.2
1.4	8.1.1	5.3.12	5.5.2, 5.8.3, 5.9.3 7.9.2,8.8.4
1.5	8.1.3	6.1	8.8.4
2.1	8.1.3, 8.2.2	6.2	8.3.2, 8.8.4
2.2	8.2.2	6.3.1	8.4.1, 8.8.4
2.3.1	4.6.2	6.3.2	8.4.3
2.3.2	4.6.3	6.3.3	8.6.2
2.3.3	4.3	6.3.4	8.5.1
2.3.4	8.4.3	6.3.5	3.2, 3.3
2.3.5	6.3.1, 6.4.1, 6.4.2	6.3.6	8.6.3
2.3.6	3.5.2	6.3.7	3.6.4
2.3.7	3.4.5	6.4	8.8.4
2.3.8	5.8.2 to 5.8.4	7.1	8.3.2
2.3.9	2.4.3, 6.6.2	8.1	8.4.1
2.3.10	3.3	8.2	8.4.2
2.3.11	3.6, 3.7	9.1	8.5.1
2.3.12	8.2.1	9.2	8.7.1
3.1	Chap 8	10.1	8.1.2, 8.5.4
3.2	8.8.2 to 8.8.4	10.2	7.16.2, 8.5.4
4.1	2.2.1, 8.2.3, 8.8.4	10.3	7.16.3, 8.5.4
4.2	8.2.3, 8.8.4	10.4	8.5.4
4.3	8.2.3	11.1	8.5.3
4.4	8.1.1	12.1	8.5.2
4.5	8.1.3	13.1	8.4.3
5.1	8.5.1, 8.8.4	14.1	8.6.1
5.2	8.8.4	14.2	8.4.4, 8.6.2
5.3.1	8.1.3, 8.2.2	15.1	8.6.3
5.3.2	8.1.3, 8.2.2	16.1	8.7.1, 8.7.2
5.3.3	8.2.2, 8.8.4	16.2	8.7.1
5.3.4	2.4.3	16.3	8.7.1
5.3.5	8.5.1	16.4	8.7.1
5.3.6	8.5.1, 8.8.4	16.5	8.7.1
5.3.7	8.5.3	16.6	8.7.1
5.3.8	8.8.5	16.7	8.7.1
5.3.9	8.8.5		

Second Edition of Guidance Note 3

References are the same as for the 1st edition except for the following

G.N.3	Guide	G.N.3	Guide
2.3	8.2	10.5	8.5.4
2.4	8.8	12.2	8.5.4
2.5	8.1.3	15.2	8.6.3
3.3	8.8	17.1	8.8.2
10.3	8.5.1	17.2	8.8.4
10.4	7.17.3, 8.5.4	17.3	8.8.3

Cross reference index for Guidance Note 4 (GN4), 1st edition
Protection against Fire

G.N.4	Guide	G.N.4	Guide
1	2.2.1, 2.2.2	3.3	3.5.3
2	3.5.2, 3.5.3, Chap 5, 7.6	4.1	3.5.2, 8.8.4
3 Gen	3.5.1	4.2	8.8.3
3.1	3.5.2, 6.5.1	4.3	8.8.4
3.1.1	3.5.2, 6.5.1, 6.5.2	5.1	2.2.5
3.1.2	3.5.2	5.2	2.2.1, 6.6.2
3.1.3	4.4.3	5.3	2.2.4
3.1.4	3.5.2	5.4	2.2.1
3.2.1	4.3.3, 4.5.1, 4.5.2	5.5	3.2.2, 6.5.2
3.2.2	4.5.2	5.6	-
3.2.3	4.3.1, 4.3.9, 6.5.1	5.7	5.1.1, 5.4.3

2nd Edition of Guidance Note 4

G.N.4	Guide	G.N.4	Guide
3.1	3.5	5.4	3.5.2
3.2.1	3.5.2, 6.5.1, 6.5.2	6.1	3.5.2
3.2.2	3.5.2	6.2	3.5.2
3.2.3	4.4.3	6.3	2.2.4
3.2.4	3.5.2	6.4	3.2.2, 6.5.2
3.2.5	3.5.2	6.5	-
3.2.6	3.5.2	6.6	5.1.1, 5.4.3
3.3.1	4.3.3, 4.5.1, 4.5.2	6.7	-
3.3.2	4.5.2	6.8	3.5.2
3.3.3	3.5.2	7.1	4.2

Cross reference Index for Guidance Note 5 (GN5), 1st Edition
Protection against electric shock

2nd Edition of Guidance Note 5

References are the same as for the 1st Editon, except for the following:

Cross reference Index for Guidance Note 6 (GN6)
Protection against Overcurrent

2nd Edition of Guidance Note 6

Note that only *changes* are listed

List of Regulations changed by the 1994 Amendments

Amended Regulations

Renumbered Regulations

List of abbreviations

A	ampere — unit of electric current	MCB	miniature circuit breaker
BC	bayonet cap	MCCB	moulded case circuit breaker
Ca	ambient temperature correction factor	MD	maximum demand
		m.i.	mineral-insulated
CENELEC	European Committee for Electrotechnical Standardisation	NICEIC	Nation Inspection Council for Electrical Installation Contracting
C_g	cable grouping correction factor		
C_i	thermal insulation correction factor	p.d.	potential difference
C_t	correction factor for the conductor operating temperature	PELV	protective extra-low voltage
		PEN	combined protective and neutral
CNE	combined neutral and earth		
cos ø	power factor (sinusoidal systems)	PME	protective multiple earthing
CPC	circuit protective conductor	PIR	passive infra-red detector
c.s.a.	cross-sectional area	PSC	prospective short-circuit current
D_e	overall cable diameter		
ECA	Electrical Contractors Association	p.v.c.	poly-vinyl chloride
EEBAD	earthed equipotential bonding & automatic disconnection	R	resistance (electrical)
		R	resistance of supplementary bonding conductor
ELCB	earth leakage circuit breaker	R_a	total resistance of protective conductor and electrode to earth
ELV	extra-low voltage		
EMC	electro-magnetic compatibility		
e.m.f.	electro-motive force	R_b	earth electrode resistance
ES	edison screw	R_p	resistance of the human body
f	frequency	RCCB	residual current circuit breaker
FELV	functional extra-low voltage	RCD	residual current device
GN	guidance note	r.m.s.	root-mean-square (effective value)
HBC	high breaking capacity (fuse)		
HRC	high rupturing capacity (fuse)	s	second — unit of time
Hz	Hertz — unit of frequency	S	conductor cross-sectional area
I	symbol for electric current	SELV	separated extra-low voltage
I_2	operating current (fuse or circuit breaker)	t	time
		TN-C	earthing system (see 5.2.5)
I_a	current to operate protective device	TN-C-S	earthing system (see 5.2.4)
Ib	design current	TN-S	earthing system (see 5.2.3)
I_d	fault current	TT	earthing system (see 5.2.2)
IEC	International Electrotechnical Commission	U	symbol for voltage (alternative for V)
IEE	Institution of Electrical Engineers	Uac	alternating voltage
I_n	current setting of protective device	Udc	direct voltage
		Uo	phase voltage
I_t	tabulated current	V	volt — unit of e.m.f. or p.d.
I_z	current carrying capacity	W	watt — unit of power
IT	earthing system (see 5.2.6)	X	reactance
k	kilo — one thousand times	Z	impedance (electrical)
kV	kilovolt (1000 V)	Z_e	earth loop impedance external to installation
L1, L2, L3	lines of a three-phase system		
m	metre	Z_s	earth fault loop impedance
m	milli — one thousandth part of	ø	phase angle
M	meg or mega — one million times	Ω	ohm — unit of resistance, reactance and impedance
mA	milliampere	µ	micro — one millionth part

205

Index

Acknowledgements

The author is indebted to the following for their help during the preparation of this Electrician's Guide:

Dr Katie Petty-Saphon, the Publisher, especially for her good humoured forbearance.

Mr N Hiller, the Editor, for his helpful advice and continued help.

Mr L. Hanner and Monsieur Xavier Flavard who cheerfully produced the figures in a very short space of time.

Mr B. Whitehouse, Publicity Manager of IMI Santon Ltd, for his help with subsection 7.11.3

Mr D. Harris, Chairman, of Robin Electronics Ltd for his help, particularly with Chapter 8.

Mr G. Peck, late of the Insurance Technical Bureau, for his advice and information.

Mr R G Cottignies, Chief Engineer of BICC Cables for help with Chapter 4.

Mr D Bailey of Huddersfield Technical College for his continued interest and help.

His wife, for her continued patience and understanding.